CALIFORNIA SAND WASPS OF THE SUBTRIBE STICTIELLINA

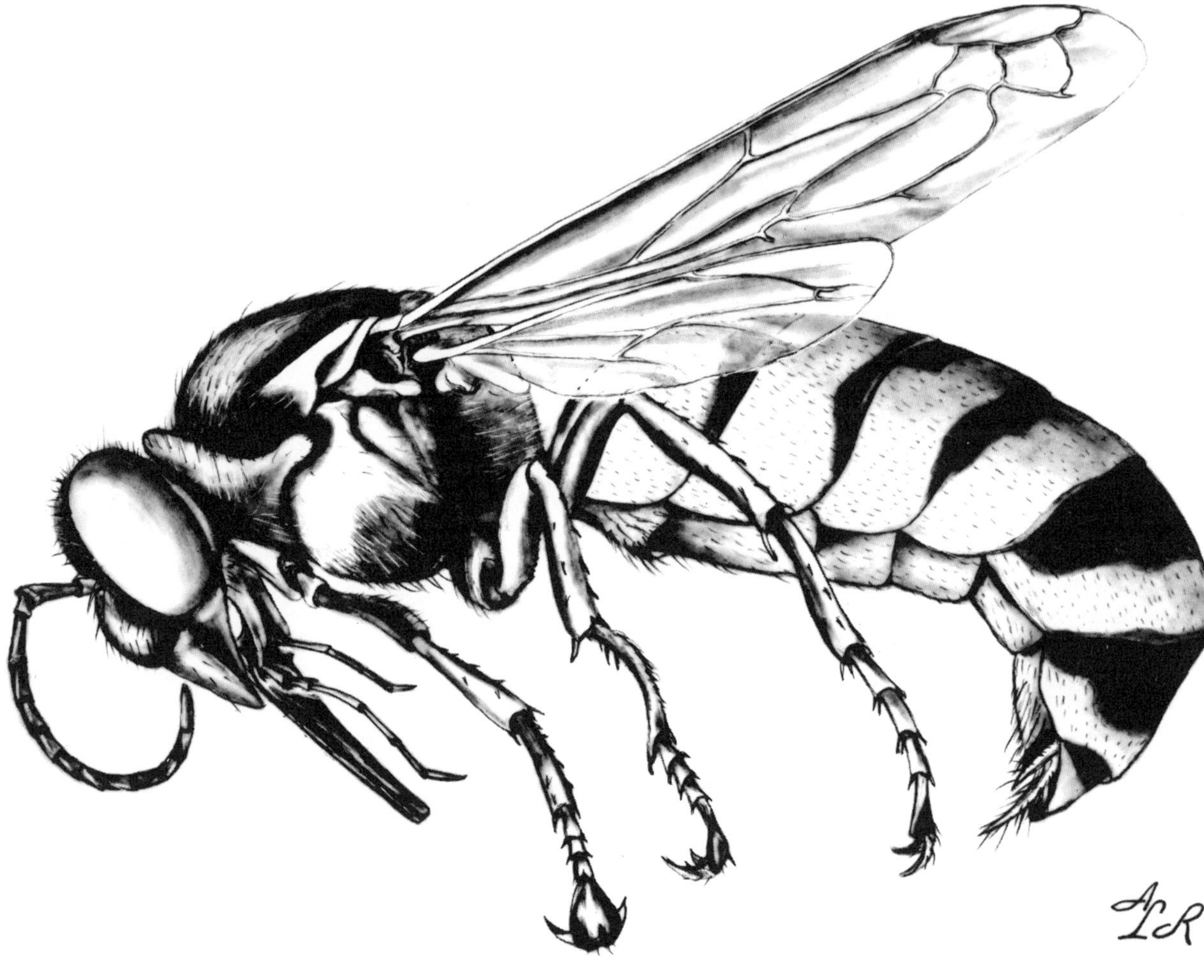

BULLETIN OF THE CALIFORNIA INSECT SURVEY
VOLUME 27

California Sand Wasps of the Subtribe Stictiellina

by R. M. Bohart and J. E. Gillaspy

UNIVERSITY OF CALIFORNIA PRESS
BERKELEY • LOS ANGELES • LONDON

BULLETIN OF THE CALIFORNIA INSECT SURVEY

VOLUME 27

Issue Date: September 1985

UNIVERSITY OF CALIFORNIA PRESS
BERKELEY AND LOS ANGELES

UNIVERSITY OF CALIFORNIA PRESS, LTD.
LONDON, ENGLAND

ISBN 0-520-09697-5
LIBRARY OF CONGRESS CATALOG CARD NUMBER 85-8565

PRINTED IN THE UNITED STATES OF AMERICA

LIBRARY OF CONGRESS CATALOGING IN PUBLICATION DATA

Bohart, R. M. (Richard Mitchell), 1913-
California sand wasps of the subtribe Stictiellina.

(Bulletin of the California Insect Survey; vol. 27)
Bibliography: p.
Includes index.
1. Sphecidae--Classification. 2. Insects--
Classification. 3. Insects--California--Classification.
I. Gillaspy, J. E. II. Title. III. Title: Stictiellina.
IV. Series: Bulletin of the California Insect Survey;
v. 27.
QL475.C3C3 vol. 27 595.709794 s 85-8565
[QL568.S7] [595.79]
ISBN 0-520-09697-5 (pbk.: alk. paper)

Contents

Acknowledgments

We appreciate the assistance, over the years, of museum curators when we have examined types or borrowed material under their care. Institutions most utilized, and abbreviated designations given in the text, are: Academy of Natural Sciences, Philadelphia (ANSP); American Museum of Natural History, New York (AMNH); California Academy of Sciences, San Francisco (CAS); California State Department of Agriculture, Sacramento (CSDA); Colorado State University, Fort Collins; Cornell University, Ithaca; Florida State Collection of Arthropods, Gainesville; Geneva (Switzerland) Natural History Museum; Iowa State University; Los Angeles County Museum (LACM); Harvard Museum of Comparative Zoology, Cambridge (MCZ); Oregon State University, Corvallis (OSU); San Diego Natural History Museum (SDM); Texas A & M University, College Station; United States National Museum of Natural History, Washington, D. C. (USNM); Universitetets Zoological Museum, Copenhagen; University of California (at Berkeley, UCB; Davis, UCD; Riverside, UCR); University of Georgia, Athens; University of Idaho, Moscow; University of Kansas, Lawrence (KU); University of Missouri, Columbia; University of Nebraska, Lincoln; University of Nevada, Reno (UNV); Utah State University, Logan; Vienna (Austria) Natural History Museum.

Museums that contribute material are usually thanked by borrowers, and rightly so. However, individuals collect the specimens, and some of these collectors who make special efforts over a considerable period should have special recognition. For many years the University of California offered summer field courses, such as Entomology 49 and its successor, 109. Many of the students and staff in these courses have been recognized previously (Gillaspy, 1964). In alphabetical order, these were: R. C. Bechtel, R. M. Bohart, J. S. Buckett, P. D. Hurd, L. S. Kimsey, R. L. Langston, E. G. Linsley, J. W. MacSwain, A. T. McClay, A. S. Menke, F. D. Parker, J. A. Powell, E. I. Schlinger, C. N. Slobodchikoff, N. J. Smith, E. C. Van Dyke, and B. Villegas. Others who have made substantial contributions to the California fauna over a period of years are G. E. Bohart, J. C. Bradley, H. E. Cott, H. K. Court, H. E. Evans, D. K. Faulkner, G. R. Ferguson, J. E. Gillaspy, T. Griswold, T. R. Haig, J. C. Hall, M. E. Irwin, R. O. Schuster, L. A. Stange, P. H. Timberlake, E. R. Tinkham, E. P. VanDuzee, and M. and J. Wasbauer. Some of these individuals were originally students in Entomology 49 or 109.

In addition to the summer field courses, there were regular California Insect Survey (CIS) spring trips organized by collectors on the University of California at Berkeley staff. These concentrated in desert areas and accumulated considerable numbers of Stictiellina. Examples are: 1952, 1958, Hopkins Well area (Riverside Co.); 1953, 1954, 1955, Borrego Valley (San Diego Co.); 1956, Cronise Valley, New York Mts. (San Bernardino Co.), and Borrego Valley (San Diego Co.); 1957, Surprise Canyon (Inyo Co.); 1959, Gorman area (Ventura Co.); 1961, Chino Canyon and Mohave area (Los Angeles Co.); 1961, Descanso area (San Diego Co.); 1963, Deep Canyon (Riverside Co.); 1964, Kernville and western Mojave Desert (Kern Co.); 1965, Lytle Creek and Elsinore (Riverside Co.); 1967, Baja California and Santa Barbara and Ventura counties; 1968, Little Rock (Los Angeles Co.); 1969, Lone Pine and Panamint Valley (Inyo Co.); 1977, Zzyzx and Kelso (San Bernardino Co.).

The frontispiece was drawn by Alyne Ruppanner; other illustrations were drawn with the aid of a camera lucida by the senior author.

CALIFORNIA SAND WASPS OF THE SUBTRIBE STICTIELLINA

INTRODUCTION

The Stictiellina comprise part of the well known sand wasps frequently called "bembicids." They represent the section of the tribe Bembicini of the Nyssoninae:Sphecidae in which the midocellus is represented by a plane recessed within a raised area. The subtribe Bembicina has been treated previously (Bohart and Horning, 1971), and this Bulletin covers the remainder of the tribe.

The stictiellins are mostly black and yellow or nearly all yellow wasps, with diversity and abundance centered in desert areas of the southwestern United States. Their conspicuous and variable markings tend to create confusing "look-alikes" in species not closely related on structural grounds. Stictiellins range from moderately small to moderately large, and at least a few specimens are found in most general collections, particularly from xeric areas of the west. They are exceedingly fast and agile fliers, and netting them can be a challenge to the collector's reflexes. Females will sting, but only if they are captured and handled roughly. Adults of both sexes commonly visit flowers, and their long tongues enable them to take nectar efficiently. Pollen is accumulated, but only through accidental adherence to body hair.

Features that are apparently adaptive with respect to fast flight and to the heat, light, wind, and dust encountered near midday, when the activity level is highest, were listed for *Steniolia* by Gillaspy (1964). Relatively stout body form may be an additional such character, providing capacious internal space to buffer extreme heat conditions.

BIOLOGY

A considerable amount of biological information is available, and it has accumulated especially during the past 20 years. These data have been contributed by J. E. Gillaspy (1951-1964), H. E. Evans (1964-1973), and J. Alcock (1975). Evans (1966) has given an excellent perspective on behavior. Here we summarize the findings of the above authors; further details are given under each species concerned. A useful list of published biological facts can be found in Krombein (1979:1716-1720).

Mating probably takes place during flight; but although the sexes have been netted together, critical observations are few. Females are sometimes present in sleeping aggregations, but there is no evidence that copulation takes place there. Males of *Glenostictia satan* cruise small territories within the colony, presumably waiting for females to emerge (Bohart, personal observation). Both sexes may be found sleeping in the above-mentioned clusters or may retire to empty burrows.

Nests occur in small to medium-sized colonies. Most of the observed colonies would fit inside a 10-meter square. The *G. satan* colony referred to above had an estimated 2,500 individuals, mostly males, present at any one time. Nests of stictiellins are shallow and are constructed in sandy soil. A closure is made when the female wasp forages for prey. Provisioning is either of the mass type, or progressive. According to Evans (1966), *Stictiella* is a mass provisioner, but in some instances the egg may hatch before provisioning is complete (delayed provisioning). In *Glenostictia* and *Steniolia* (Evans, 1966) as well as *Xerostictia* (Alcock, 1975), only the presumably more evolutionarily advanced progressive sort is known. In other words, provisions are placed in the open cell successively as the larva develops. This contrasts with the more usual method in predatory Hymenoptera, in which all provisions are collected and the cell closed before the wasp egg hatches (mass provisioning). The egg is placed in the empty cell in *G. scitula*, but on the side of the first provision in other *Glenostictia* and in *Steniolia* and *Stictiella*.

Stictiella and *Microstictia* prey on Lepidoptera, both butterflies and moths, especially small specimens. Nests in these genera may be multicellular. *Steniolia* and *Glenostictia* stock their unicellular nests with flies, especially bombyliids. An exception is *G. scitula*, which takes various small Hymenoptera, particularly bees, as well as Hemiptera (including Homoptera) and flies (Gillaspy et al., 1962). In *Xerostictia*, prey have been reported by Alcock (1975) to be adult antlions (Myrmeleontidae) and Homoptera (Flatidae).

Many species, particularly those with relatively short wings, produce a high-pitched "whining" sound during flight. This may be particularly noticeable when they are hovering and is not unlike the noise produced by some of the Bombyliidae. It is unquestionably an aid to the collector, who can spot the wasp after first hearing the sound. In the case of a large colony such as that of *G. satan*, a pronounced hum can be detected at a distance of 10 meters.

Parasite relationships have not been extensively investigated. Miltogrammine flies are known to be active around nests, but few instances of actual parasitization are known. Mutillids undoubtedly reduce populations to some extent, as they do in other sphecids. The chrysidid *Parnopes* may be an important predator at times, especially *P. edwardsii* Cresson and other large species of the genus (Evans, 1966). Habits of *Parnopes* and a key to the American species were given by Bohart and Kimsey 1982.

Studies of stictiellin larvae have been made by Evans and Lin (1956) and Evans (1964). Species described were *Glenostictia pulla* and *scitula*; *Stictiella pulchella*; and *Steniolia elegans*, *nigripes*, *duplicata*, and *obliqua*. A key to the species based on larvae was given by Evans (1964).

Flower visitation records are few in the literature, and most of them have been taken from data on pinned specimens. No differentiation is made in these cases between visits for nectar, to seize prey, or in males for mating with females attracted to flowers. In any case, plant genera that occur repeatedly are: *Eriogonum*, *Petalonyx*, *Prosopis*, *Baccharis*, *Chrysothamnus*, *Condalia*, *Hyptis*, *Heliotropium*, *Tamarix*, *Melilotus*, *Asclepias*, and *Croton*.

DISTRIBUTION

Steniolia are relatively abundant in forested areas of the west, and most general collections from the Sierra Nevada, Cascade Range, and Rocky Mountains contain some of this genus. Arid regions of the southwest seem to be preferred by some *Steniolia* and nearly all of the other Stictiellina.

Species of the subtribe are strictly New World inhabitants, and a study of their geographical distribution reveals some interesting statistics:

1. Only 5 species occur east of the Mississippi River: *Glenostictia emarginata*, *G. pictifrons*, *Stictiella pulchella*, *Microstictia minutula* (Handlirsch), and *M. texensis* Gillaspy. Only the last two have their ranges predominantly east of the 100th meridian.
2. Of the 61 known species of the subtribe, 37 occur in California, and most of these are centered in the southeastern corner or in arid parts of the Owens Valley and associated mountain ranges to the east.
3. Many species occur in Mexico, including Baja California.
4. Only 2 widespread species have bridged the Panama Canal and have been collected in Ecuador and Venezuela, respectively: *Steniolia guatemalensis* (Rohwer) and *S. longirostra* Say.

In distribution records, California counties are arranged approximately from north to south. Full data are given for newly described and rare forms.

PHYLOGENY

The Bembicini, with their unusually compact and streamlined form, are generally considered to be among the more specialized tribes of the Sphecidae (Bohart and Menke, 1976). The most obvious characteristic that separates the tribe is the lengthened labrum (at least as long as broad). Characters within the tribe that serve as a measure of specialization are the reduction in number of palpal segments and reduction toward obsolescence of the ocelli. These tendencies are well marked, both in subtribe Bembicina and in Stictiellina. In the latter subtribe, the conventional sphecid formula of maxillary and labial palpal segments of 6-4 becomes progressively 5-2, 4-2, 3-2, and 3-1 among the various genera. However, this is not a straight-line phenomenon, as shown in Table 1 and Figure 1.

Other tendencies are the narrowing of the ocellar remnants, peaking of the clypeus between the antennal bases, reduction in the arolium, lengthening (Chilostictia[1], Xerostictia) or strengthening (Steniolia) of the labrum, lengthening of the galea and other tongue elements, and distortion of the male midfemur. Indeed, some species of Glenostictia exhibit all 11 ancestral character states. Of the 6 genera recognized in Stictiellina, none combines all of the derived features, but most of them are found in Steniolia and Xerostictia, and the fewest are in Microstictia and Glenostictia. The outgroup on which the ancestral character states in Table 1 are predicated is the genus Stizus of the tribe Stizini. In this tribe the mouthparts are relatively short and the labrum, though prominent, is always broader than long.

Two tendencies that are particularly intriguing and stimulate speculation as to evolutionary significance are the modifications of the midocellar plane and the reduction of the arolia, particularly in the female. It can be surmised that the midocellar plane, containing remnants of a lens, was at one time and even now may be marginally light sensitive. Its depression below the surrounding surface, and the development of partly encircling ridges, is likely a protective adaptation related to the strong rays of the sun in desert habitats. It is interesting that this "protection," if it is indeed that, took place in Glenostictia, Chilostictia, and Steniolia through a narrowing of the plane (fig. 6), and in Microstictia through its broadening and shortening (fig. 21).

No tendency toward reduction of the arolia occurs in Glenostictia, Chilostictia, and Steniolia, all predators on flies. In these genera a substantial adhesive pad in the female would seem to be an obvious advantage. In the male, the arolia may well have a holding function during mating. However, for Microstictia and Stictiella the predominant prey, as far as we know, are moths and sometimes butterflies. Here the female arolia are reduced or even absent, with a concomitant increase in claw size (fig. 17). It seems logical that a shorter arolium and longer claws are better fitted to deal with scaled surfaces in Lepidoptera adults. At the same time, arolia of males in these two genera (except in the Stictiella pulchella group) are not significantly reduced, and presumably retain a function during mating.

[1]Chilostictia Gillaspy 1983:286, generotype: C. hirsuta Gillaspy, Baja California Sur.

STRUCTURAL CHARACTERS

There are many structural features that vary from species to species, and these form the basis of the keys and descriptions. Coloration is a less reliable taxonomic character, but it has an average or modal value, and the pattern to which it contributes is often useful. Females nearly always have a larger proportion of white or yellow markings than males of the same species. An important exception is Steniola sulfurea, in

Table I: Character States

The following character states are those which seem to be evolutionary in nature, and they are indicated by number on the dendrogram (Fig. 1). Characters which occur in some species of a genus but not others are identified with the ± sign.

Presumably unspecialized

1. Palpal formula 6-4
2. Labrum about as long as broad or a little longer than broad
3. Labrum simply convex
4. Galea not longer than eye height
5. Clypeal margin straight or concave above
6. Midocellar plane as broad as long
7. Midocellar plane not much, if any, broader than long
8. Foreleg arolium developed and at least half as long as tarsal V
9. Foreleg arolium present
10. Midfemur relatively simple (rarely notched distally)
11. Midfemur not distorted

Presumably specialized

1a. Palpal formula 5-2
1b. Palpal formula 4-2
1c. Palpal formula 3-2 or 3-1
2a. Labrum considerably longer than broad
3a. Labrum humped toward base
4a. Galea longer than eye height
5a. Clypeal margin peaked above
6a. Midocellar plane longer than broad
7a. Midocellar plane greatly broadened
8a. Foreleg arolium (if developed) less than half as long as tarsal V (in female at least)
9a. Foreleg arolium undeveloped
10a. Midfemur serrate or irregularly carinate but not otherwise distorted
11a. Midfemur distorted

which the male abdomen is almost all yellow. Size in terms of body length, robustness, and wing length are species characteristics, but may vary considerably, depending presumably on amount of larval nutrition. Also, in many species males average slightly larger than females. The nature and location of pubescence, particularly on the clypeus and hindfemur, are useful taxonomic criteria. Clypeal "silvering" seems to have a heat-reflecting significance and is accentuated in desert species. Erect and often black hair on legs and other body parts is most abundant in species found in montane situations, where it may have a thermoregulatory function.

Other structural characters are more constant within species but vary between species, and therefore are more useful taxonomically. These are clypeal and labral shape, length of the galea, form of the flagellomeres and their relative lengths, breadth of the frons compared with the eye, degree of narrowing or broadening of the midocellar plane, shape and setation of tarsal elements, serration or other modifications of the male midfemur, contour (as well as pubescence) of the hindfemur, shape of male tergum VII including development of its ventral (normally hidden) spiracular lobes, presence and development of projections on male sternum II, posterior outline of male sternum VI, number of projections on male sternum VIII, and particular formation of male genitalia. Of special importance in Steniolia males is the nature of the shallow oval depressions ("fossulae") on the underside of the antennae (figs. 28h, 29h). These are associated with welts or tyloides, and they should be viewed with good light at a magnification of 25-50 diameters.

TAXONOMIC METHODS

In keys and descriptions, certain abbreviations have been used in the interest of brevity. These are: F-I, F-II, etc., for antennal flagellomeres; LID for least interocular distance; T-I, T-II, etc., for abdominal terga; and S-I, S-II, etc., for sterna. Abdominal segments are numbered from the base without regard to propodeum. Abbreviations for southern, northern, etc., are given as s. and n. to avoid confusion with New Mexico, South Dakota, etc.

In order to see important characters in the male, it is necessary to extend the last few abdominal segments and extract the genitalia. This will reveal both S-VII and its overlapping spiracular lobes of T-VII, and S-VIII with its 3-4 projections. While specimens of both sexes are still fresh, it is advisable to spread the mandibles and extend the tongue far enough to expose the palpi and afford a clear view of the galea.

NOMENCLATURE

Species in the subtribe Stictiellina were placed under a variety of generic names by workers between 1850 and 1930. F. Smith (1856) used Monedula Latreille (a synonym of Stictia Illiger) for his pictifrons. Cresson (1865) and W. Fox (1895) recognized Steniolia Say, but all others were put in Monedula. Parker (1917, 1929) and C. Fox (1923) differentiated Steniolia but lumped the rest under Stictiella J. Parker. Since 1951, a series of papers by Gillaspy have divided Stictiella into the presently recognized Stictiella, Glenostictia, Xerostictia, Microstictia, and Chilostictia.

SYSTEMATICS

KEY TO THE GENERA OF STICTIELLINA

1. 11 antennal flagellomeres, abdomen with 7 terga (males). 2

   10 antennal flagellomeres, abdomen with 6 terga (females). 7

2. Labrum strongly bulging toward base, galea more than 2x as long as labrum, midfemur entire ventrally. . . . . .Steniolia Say

   Labrum not bulging toward base, galea less than 2x as long as labrum, midfemur sometimes serrate ventrally or otherwise modified. 3

3. Labrum 1.7x as long as broad or longer, palpal formula less than 6-4. 4

   Labrum less than 1.5x as long as broad, palpal formula 6-4. 5

4. Midfemur entire beneath, labrum 1.7x as long as broad, palpal formula 5-3, hindfemur with abundant long hair ventrally . . . . .Chilostictia Gillaspy

   Midfemur serrate beneath, labrum 2x as long as broad, palpal formula 4-2, hindfemur with some short erect hairs ventrally. . . . . Xerostictia Gillaspy

5. Midocellar plane about 2x as long as broad, midfemur sometimes irregular but not serrate. . . . . . Microstictia Gillaspy

Midocellar plane less than 2x as broad as long, midfemur entire, distally notched, or serrate beneath, not otherwise deformed. 6

6. Midfemur entire beneath, not carinate; one species with a large distal notch; upper clypeal margin sometimes peaked medially; midocellar plane sometimes distinctly longer than broad. . . . . .Glenostictia Gillaspy

Midfemur serrate beneath or with a carina on distal three-fourths which ends in a small notch; upper clypeal margin straight or concave medially; midocellar plane about as long as broad or a little broader. . Stictiella J. Parker

7. Labrum bulging toward base, galea more than 2x as long as labrum . . . Steniolia Say

Labrum not bulging toward base, galea less than 2x as long as labrum. 8

8. Labrum 1.7x as long as broad or longer, palpal formula less than 6-4. 9

Labrum less than 1.5x as long as broad, palpal formula 6-4. 10

9. Labrum 1.7x as long as broad, palpal formula 5-3, hindfemur with much erect hair posteroventrally..Chilostictia Gillaspy

Labrum 2x as long as broad, palpal formula 4-2, hindfemur with few very short, erect hairs posteroventrally . . . . Xerostictia Gillaspy

10. Midocellar plane about 2x as broad as long (fig. 26b). . . . .Microstictia Gillaspy

Midocellar plane less than 2x as broad as long. 11

11. Foretarsal arolium about half as long as claws (fig. 7f, g); clypeus not strongly convex, margin sometimes reaching a peak at middle above; midocellar plane sometimes distinctly longer than broad . . . . .Glenostictia Gillaspy

Foretarsal arolium at most one-third as long as claws, sometimes apparently absent (figs. 23f, g, 16f, g); clypeus rather evenly and strongly convex, margin straight or concave at middle above; midocellar plane as broad as long or a little broader. . . Stictiella J. Parker

Genus Glenostictia Gillaspy

Bembix, Monedula, Stictiella of authors in part.
Glenostictia Gillaspy 1962:563 (in Gillaspy et al.). Generotype: Monedula pulla Handlirsch, original designation.

The genus contains 21 nearctic species, and most of these are limited to the area west of the 100th meridian. Glenostictia is well represented in the California fauna, with 13 species. Several species extend their range into Mexico, and 2 have been found only there.

Biology. Information has been published for 8 of the species. Females mass-provision their ground nests with a variety of flies, particularly Bombyliidae. Other prey families noted in the literature are Tachinidae, Sarcophagidae, Muscidae, Therevidae, Apioceratidae, and Syrphidae. One species, G. scitula, has rather catholic tastes in prey, which include a variety of flies, wasps, ants, and bees, as well as leafhoppers. More details on prey are given under individual species.

Nests may occur in a small, loose aggregation or may form a large, tight colony. One of the latter was observed by R. M. Bohart for G. satan in southwestern New Mexico. The nesting site was roughly circular, with a diameter of about 10 meters. More than 2,000 males and an undetermined number of females cruised over this area, producing a loud, high-frequency hum which could be heard at a distance of about 10 meters. Apparently this site was later abandoned, because there was no sign of the wasps the following year.

Taxonomy. The most diagnostic morphological characters of Glenostictia are: (1) midocellar plane as long as broad or longer; (2) labrum nearly flat in lateral profile; (3) palpal formula 6-4; (4) clypeus gently convex in lateral profile; (5) arolium distinct between moderate-sized claws; (6) clypeus sometimes projecting upward between antennal bases dorsally; (7) male midfemur smooth and rounded beneath, rather than serrate or sharply carinate (tenuicornis has a ventrodistal notch); (8) male S-VIII with a ventral tooth (only a carina in clypeata).

The other genera with a 6-4 palpal formula are Stictiella and Microstictia. Males of Glenostictia are readily differentiated from Stictiella by the absence of a carina or serration beneath the midfemur (character 7 above), and from Microstictia by the longer midocellar plane (character 1 above). Females can also be separated from Microstictia by the midocellar shape (compare figs. 9b and 24b). However, differences from Stictiella are less obvious. Some Stictiella can be removed from consideration by the practical absence of an arolium between unusually long and slender claws (pulchella group). Similarly, some Glenostictia can be eliminated because of a clypeus which protrudes upward between the antennal bases (pulla and scitula groups). G. bifurcata, clypeata, tenuicornis, megacera, and bituberculata have the midocellar plane considerably longer than broad (instead of as broad or broader than long), and in addition the first four have the clypeus angled up medially above, as does satan. The remaining Glenostictia are the pictifrons group (pictifrons, arizonae, mexicana, nigriloba, vechti, veracruzae), and terlinguae, parva, and californica. These have in common with Stictiella a flat or concave upper clypeal margin and an unlengthened midocellar plane. In the pictifrons group, as well as parva, californica, and terlinguae, the clypeus is much less bulging than in Stictiella, the foreleg arolium of females extends more than one-third of the claw length in dorsal or end view (one-fourth or less in Stictiella), and there is a yellow band across the upper two-fifths to one-fourth of the female clypeus in the pictifrons group.

The separation of Glenostictia into groups is difficult. Numbers 1-3 below are relatively homogeneous, but the remaining 8 species do not seem to have any close relatives.

1. The pictifrons group. Midocellar plane not much if any longer than broad; clypeus not peaked above; female clypeus convex and with a transverse yellowing above; male foretarsals II-IV somewhat broadened, as well as flattened and usually lobate; male midbasitarsus curved; male S-II not dentate; female foretarsal V not bristly beneath. Included species: arizonae, mexicana, nigriloba, pictifrons, vechti, veracruzae.
2. The pulla group. Midocellar plane longer than broad; clypeus peaked above and extending partly between antennal sockets; female clypeus somewhat flattened on either side of middle; male foretarsus simple; male midbasitarsus straight; male S-II bidentate; female foretarsal V bristly beneath. Included species: angulata, angulifera, argentata, gilva, pulla.
3. The tenuicornis group. Midocellar plane longer than broad; clypeus a little peaked above, female clypeus flattened on either side of middle; male tarsi unmodified; male S-II not dentate; female foretarsal V not bristly beneath. Included species: clypeata, tenuicornis.
4. Miscellaneous species. G. bifurcata, bituberculata, californica, megacera, parva, satan, scitula, terlinguae.

KEY TO SPECIES OF GLENOSTICTIA

1. Males, 7 visible terga, 11 flagellomeres 2

   Females,[1] 6 visible terga, 10 flagellomeres 22

2. Hindfemur beneath with considerable short to long hair, midbasitarsus essentially straight 3

   Hindfemur beneath with at most a few short, erect hairs; midbasitarsus various. . 5

3. Labrum with much long, erect hair; S-II bidentate (w. U.S., n. Mexico) . . . . scitula (W. Fox)

   Labrum without long, erect hair; S-II with a median projection. 4

4. S-II projection simple; flagellum thickened toward base, F-I not longer than F-IX-XI together (w. U.S. to Colorado, n. Mexico). . . . . megacera (J. Parker)

   S-II projection bifurcate; flagellum slender toward base, F-I longer than F-IX-XI together (sw. US, nw. Mexico). . . . . bifurcata (C. Fox)

5. Clypeus with dorsomedial margin angled and reaching up well above imaginary line drawn across bottom of antennal sockets, midbasitarsus essentially straight. 6

   Clypeus either nearly straight-across dorsomedially or angled up somewhat, but

[1]Females of mexicana are unknown.

not above imaginary line drawn across bottom of antennal sockets; midbasitarsus various. 10

6. Scutum densely silvery, obscuring punctation; T-VII spiracular lobes wedgelike in posterior view, broad, apically emarginate, nearly touching medioventrally (s. Utah to California, nw. Mexico) . . . . argentata (C. Fox)

Scutum partly silvery but not obscuring punctation; T-VII spiracular lobe thin in posterior view, narrow or broad, sometimes weakly emarginate. 7

7. S-VI nearly straight to slightly rounded posteriorly, T-VII spiracular lobes narrow and well separated (fig. 12c). .8

S-VI angled back posteriorly, T-VII spiracular lobes broad and nearly touching medioventrally (fig. 2c). . . 9

8. Femora black-marked in front as well as behind, clypeus nearly always half to four-fifths dark, flagellum mostly dark beyond F-I or F-II, terga whitish toward middle at least (nw. U.S., Baja California Mts.). . .pulla (Handlirsch)

Femora yellow in front, clypeus all pale or with only a little darkening laterally, flagellum mostly orange beyond F-II, terga various (w. U.S. to w. Texas, n. Mexico) . . . . gilva Gillaspy

9. Flagellum dark brown beyond F-II or F-III, clypeus usually broadly black (fig. 2a, d) (s. California, Baja California). . . . . .angulata Gillaspy

Flagellum pale reddish beneath from F-II to F-VII or more, clypeus nearly all yellow or narrowly black-margined laterally (fig. 3a, d) (s. California, Nevada) . . . . .angulifera Bohart

10. Midbasitarsus essentially straight overall, a little incurved basally but without a clump of bristles on concave side, foretarsus simple. 11

Midbasitarsus distinctly curved and bearing a clump of bristles near middle of concave side. 16

11. F-I shorter than F-IX-X together, S-II with a median toothlike or spinelike projection, T-I-III (at least) with free submedian yellow spots (Texas to Arizona). . . . . . .terlinguae (C. Fox)

F-1 longer than F-IX-X together, S-II without projections or with a pair of bumps or teeth, T-III (at least) without free yellow spots. 12

12. F-XI about 2.5x as long as broad (fig. 7a), S-II often with a medioposterior pair of bumps or teeth (California, Arizona, nw. Mexico) . . . . .bituberculata (J. Parker)

F-XI no more than 2x as long as broad, S-II without toothlike projections. .13

13. Thoracic dorsum all black, rest of body largely black (Arizona, New Mexico). . . . . satan Gillaspy

Thoracic dorsum marked with yellow or whitish. 14

14. Midfemur with a large distal notch beneath (fig. 14o), S-VIII with a median tooth (California, Nevada, Arizona, Baja California). . . . .tenuicornis (W. Fox)

Midfemur entire; S-VIII various. 15

15. Facial markings yellow, spiracular lobes of T-VII narrow (fig. 9c), S-VIII without a medioventral tooth (fig. 9m), body length 15 mm or more (s. California to to Texas, n. Mexico) . . . . .clypeata (Gillaspy)

Facial markings whitish, spiracular lobes of T-VII broad (fig. 8c), S-VIII with a medioventral tooth (fig. 8m), body length less than 15 mm (se. California, sw. Arizona). . . . . californica Bohart

16. Foretarsals III-IV all pale beneath, small species 9-11 mm long (Arizona to w. Texas, nw. Mexico). . . . . parva Bohart

Foretarsals III-IV black-marked beneath, medium-sized to large species, more than 11 mm long. 17

17. Foretarsomere II not unusually expanded toward posterodistal corner, at most narrowly margined with black dorsoposteriorly, S-II more than half black. .18

Foretarsomere II expanded toward posterodistal corner or with a large black dorsal spot, S-II less than half black 19

18. Foretarsomere V but not II black-margined posteriorly, lateral lobes of T-VI well separated ventrally, median projection of S-VIII spiniform (Veracruz, Jalisco, Oaxaca). . . . . veracruzae Bohart

Foretarsomere II but not V black-margined posteriorly, lateral lobes of T-VI overlapping ventrally, median projection of S-VIII an obtuse tooth (Jalisco). . . . . . . .mexicana Bohart

19. Foretarsomere II dorsally with a black lobe posterodistally, I sometimes black-spotted. 20

Foretarsomere II without a black lobe, I and II sometimes with discolored posterior edge. 21

20. Clypeus yellow, hindtarsomeres II-IV black maculate dorsally; last antennal article (F-XI) stout and distinctly curved at apex; lobes of foretarsomeres III-IV rather pointed (U.S. w. to Arizona, Mexico). . . pictifrons (F. Smith)

Clypeus whitish, hindtarsomeres III-IV without black marks dorsally, sometimes all hindtarsomeres pale; last antennal article (F-XI) slender and only slightly curved overall; lobes of foretarsomeres III-IV rounded distally (Arizona to w. Texas, n. Mexico). . vechti Bohart

21. Foretarsomeres III-IV with bulbous lobes, V less than 2x as long as broad, I about 3x as long as broad at apex and with rake setae thickened distad; F-XI slender and moderately incurved; S-II somewhat swollen subposteriorly (Arizona, New Mexico, n. Mexico) . . . . nigriloba Bohart

Foretarsomeres III-IV with small and tapering lobes (fig. 5m), V about 2x as long as broad, I 4-5x as long as broad at apex and with rake setae slender and tapering; F-XI stout and strongly incurved (fig. 5a); S-II not swollen subposteriorly (s. California to w. New Mexico, nw. Mexico) . . . . arizonae Bohart

22. Clypeus with dorsomedial margin clearly angling above an imaginary line drawn across bottom of antennal sockets. . 23

Clypeus with dorsomedial margin below or just reaching imaginary line drawn across bottom of antennal sockets. . .28

23. Hindfemur with conspicuous erect hair beneath, tarsomere V of each leg not bristly beneath (w. U.S., n. Mexico) . . . . scitula (W. Fox)

Hindfemur practically bare beneath, tarsomere V of each leg bristly beneath .24

24. Scutum with dense silvery appressed pubescence obscuring punctation (s. Utah to California, nw. Mexico) . . . . .argentata (C. Fox)

Scutum with moderate pubescence not obscuring punctation. 25

25. Clypeus with little appressed silvery pubescence, surface sparsely striatopunctate (s. California, Baja California). . . . . .angulata Gillaspy

Clypeus with dense and silvery appressed pubescence. 26

26. Forebasitarsus with 3 well developed preapical rake setae and a weak basal one; black basal area of T-II simple, or produced backwards medially into a pair of rather narrow projections, or with small and isolated black spots (as in fig. 3e) (s. California, Nevada) . . . . .anguliféra Bohart

Forebasitarsus with 4-5 well developed preapical rake setae and a weak basal one, black area of T-II produced medially into a broad truncation or 2 stout projections (as in figs. 10e, 12e). . 27

27. Scutum with discal spots small or absent, femora often with black marks in front as well as behind, terga yellow (nw. U.S., Baja California Mts.). . . . . pulla (Handlirsch)

Scutum with discal spots prominent or terga whitish, femora all yellow in front (w. U.S. to w. Texas, n. Mexico) . . . . gilva Gillaspy

28. Hindfemur with numerous but scattered erect hairs on outer ventral surface, midocellar plane long oval, black mark of T-I 3-pronged, clypeus angled above and just reaching line drawn across

lower edge of antennal sockets (sw. U.S., nw. Mexico) . . bifurcata (C. Fox)

Hindfemur with at most a few short erect hairs, other characters various. . . .29

29. Clypeus with a large patch of semierect, discal hair on either side of middle; midocellar plane long oval; no posterior yellow scutal spot (w. U.S. to Colorado, n. Mexico) . .megacera (J. Parker)

Clypeus with pubescence inconspicuous, or with silvery appressed hair; midocellar plane and posterior scutal spotting various. 30

30. Clypeus on either side of middle, with definite flattened areas with silvery appressed pubescence; midocellar plane long oval. 31

Clypeus rather evenly convex, midocellar plane various. 32

31. T-II sublaterally with extremely fine and close punctation, finer than on T-III; black marking toward middle of T-II usually separating yellow spots (s. California to Texas, n. Mexico). . . . . .clypeata Gillaspy

T-II sublaterally with punctation more coarse, about like that on T-III; black marking toward middle of T-II forming 2 transverse spots or a bar (California, Nevada, Arizona, Baja California). . . . . tenuicornis (W. Fox)

32. Scutum with posterodiscal yellow spot, sometimes a complete yellow U; midocellar plane various. 33

Scutum without posterodiscal yellow spot (or only a tiny one), plane of midocellus about as broad as long. 34

33. Scutum with anterodiscal yellow spots tapering in front, T-I-III usually not all with free yellow spots (California, Arizona, nw. Mexico) . . . . .bituberculata (J. Parker)

Scutum with anterodiscal yellow spots blunt in front, T-I-III with free yellow spots (Texas to Arizona) . . . . .terlinguae (C. Fox)

34. Scutum all black, T-I mostly black, tergal markings forming isolated pale spots (Arizona, New Mexico). . satan Gillaspy

Scutum not all black, at least with lateral pale spots. 35

35. Trochanters mostly black, sterna extensively black (Veracruz, Jalisco, Oaxaca) . . . . .veracruzae Bohart

Trochanters mostly yellow, sterna extensively yellow. 36

36. Forebasitarsus about 2.5-2.7x as long as broad (Arizona, New Mexico, n. Mexico) . . . . nigriloba Bohart

Forebasitarsus at least 3x as long as broad. 37

37. Clypeus all whitish, basal segment of labial palpus all pale, LID not greater than eye breadth. 38

Clypeus yellowed at least across base, basal segment of labial palpus maculate, LID usually greater than eye breadth 39

38. T-I-IV whitish medially (vernal species) (se. California, s. Nevada, nw. Arizona) . . . . californica Bohart

T-I-IV yellow medially, or rarely whitish on I-II only (autumnal species) (Arizona to w. Texas, nw. Mexico). . . . . parva Bohart

39. S-II basal black area with sharp medial extension reaching to basal third of segment; T-II with pale yellow cross mark which is broken submedially, or medial mark narrowly attached, basal black area subrectangular; mesopleural signum usually within large black mark (s. California to w. New Mexico, nw. Mexico). . . . . arizonae Bohart

S-II basal area hardly evident, not sharply extended medially, T-II with yellow band; mesopleural signum in yellow or small black area. 40

40. T-II black mark usually with pedicel which is no more than half as broad as widest point (U.S. w. to Arizona, Mexico) . . . . .pictifrons (F. Smith)

T-II black mark usually with pedicel which

is two-thirds as broad as widest point (Arizona to w. Texas, n. Mexico) . . . . *vechti* Bohart

Glenostictia angulata Gillaspy, (new species) (Fig. 2; map 1)

Male holotype. Length 13 mm, forewing 9 mm. Black, extensively marked with yellow as follows: face (fig. 2d), narrow postorbital band ending at inner eye angle, pronotum narrowly and pronotal lobes posteriorly, scutum laterally, large scutellar spots, metanotum, large mesopleural bands separated ventrally, metapleuron mostly, propodeal spots in enclosure and on posterolateral angles, legs except linear marks on femora above and below, broad tergal bands (fig. 2e), sterna posteriorly. Scape, pedicel, and F-I in front, F-II-XI entirely dark brown. Pubescence pale, that on clypeus, lower frons, and mesopleuron appressed silvery, some erect hair on clypeus; hindfemur without erect hair ventrally. Punctation inconspicuous, minute on clypeus and labrum. Facial proportions as in Figure 2d, galea 0.8 length of eye height and 1.5x length of labrum, upper clypeus margin extending between antennal sockets, F-XI 2x as long as broad, clypeus weakly but evenly convex, foretarsomere I 5x as long as broad and with 5 short rake setae; midbasitarsus straight, T-VII weakly notched at apex, spiracular lobes broad and nearly touching ventrally (fig. 2c), S-II with a pair of thornlike projections (fig. 2f), S-VI posterior margin angled backward medially (fig. 2k), S-VIII with a strong medioventral projection (fig. 2i), gonostylus moderately broad and not hairy beneath (fig. 2g, h).

Female. Length 10-13 mm. Markings similar to male, but yellow somewhat more extensive; pattern of mesonotum and metanotum as in Figure 2b, T-II-III bands as in Figure 2f. Clypeus nearly bare, whitish, with a few longitudinally oriented punctures and associated erect hairs. Forebasitarsus 3x as long as broad, with 3-4 well developed preapical rake setae; tarsomere V with numerous stout, erect setae beneath.

Geographic range. Upper Sonoran Life Zone of s. California and Baja California (map 1). Its range is similar to that of *angulifera*, but specimens have often been taken at somewhat higher altitudes.

Map 1. California distribution of *Glenostictia angulata* Gillaspy. Inset: overall distribution.

Records. Holotype male, Tanbark Flat, Los Angeles Co., California, VI-20-50 (R. O. Schuster, UCD). Paratypes, 28 males, 32 females. *California*: INYO CO.: Lone Pine Canyon, VII-4-33 (P. Timberlake); San Benito Co.: Idria, VI-8-81 (N. Smith); San Luis Obispo Co.: La Panza, VIII-7-55 (J. Powell); Creston, VII-31-77 (L. Guenther); SAN BERNARDINO CO.: Cajon Pass (or Junction), VI-24-49 (H. Cott), VIII-25-53 (J. Hall), VIII-11-56 (E. Schlinger, J. Hall), VI-7-58 (E. Schlinger); Camp Baldy, VII-11-50 (K. Whitesell); Deep Creek, VI-12-64 (R. Snelling); near Wrightwood, VI-17-56 (G. Ferguson, VII-1-56 (G. Stage); New York Mts., VI-12-79 (Brown and D. Faulkner); KERN CO.: Greenhorn Mts., V-26-46 (E. Van Dyke); Cameron Road, VIII-3-82 (W. Tyson); SANTA BARBARA CO.: San Rafael Mts., VI-29-59 (R. Bohart, F. Parker); VENTURA CO.: Sespe Canyon, VII-10-59 (J. Russell, C. Campbell); LOS ANGELES CO.: Crystal Lake, VI-29-50 (H. Robinson); Elizabeth Canyon, VII-7-59 (E. Schlinger); Tanbark Flat, VI-20 to VI-30-50 (A. McClay, T. Haig, H. Robinson, R. Schuster, P. Hurd, K. Whitesell, F. Williams), VII-10 to VII-30-50 (J. Hall, K. Whitesell, R. Schuster), VI-20 to VII-12-52 (R. Anderson, E. Evans, J. Gillaspy, D. Thompson), VI-18 to VI-30-56 (B. Bartosh, R. Bohart, J. Hall, A. Menke, H. Moffitt, E. Schlinger), VII-9-60 (M. Stebbins); Big Dalton Dam, VII-15-52 (E. Schlinger); RIVERSIDE CO.: Idyllwild, VII-14-12 (P. Timberlake); 5 mi w. Sage, VII-2-63 (E. Schlinger); SAN DIEGO CO.: near Warner Springs, VIII-8-51 (J. Gillaspy), VII-4-56 (L. Stange et al.). Mexico: BAJA CALIFORNIA: San Vicente, VII-8-63 (C. Frady).

Discussion. *Glenostictia angulata* is closely related to *angulifera*, as pointed out under that species, differing in the male mainly by the darker flagellum and clypeus. The female is likewise less extensively yellow than in *angulifera*, and the clypeus is strikingly different. It is almost bare of appressed silvery pubescence and has only a few widely separated striatiform punctures. In *angulifera*, the surface is covered by appressed silvery pubescence, obscuring punctation.

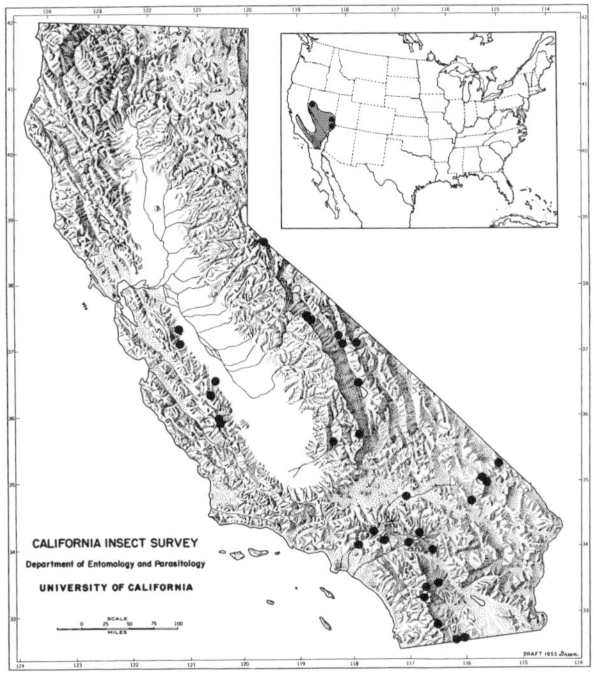

Map 2. California distribution of *Glenostictia angulifera* Bohart. Inset: overall distribution.

Glenostictia angulifera Bohart, (new species) (Fig. 3; map 2)

Male holotype. Length 12 mm, forewing 9 mm. Black, extensively marked with yellow as follows: face about as in Figure 3d, postorbital streak broken at inner eye angle, pronotum mostly, scutum laterally, stout discal spots, large scutellar spots, metanotum, large mesopleural spot joined ventrally but not covering signum, metapleuron except posteriorly, a pair of nearly joined transverse spots on propodeal enclosure, rest of propodeum except large medial triangle, legs except spots behind femora and trochanters, terga posteriorly (fig. 2e), sterna posteriorly, gonostylus at apex. Scape, pedicel, and F-I yellow in front, F-II-VII light reddish in front, rest brownish. Pubescence pale, appressed and silvery on clypeus, labrum, and lower frons, scutum anteriorly, and mesopleuron; some erect hair on upper half of clypeus; hindfemur without erect hair ventrally. Punctation inconspicuous, minute on clypeus and mostly so on labrum. Facial proportions as in Figure 3d, galea a little shorter than head height, 1.5x as long as labrum, upper clypeal margin extending upward halfway between antennal sockets, F-XI 2x as long as broad, clypeus weakly but evenly convex, midocellar flat area 1.5x as long as broad. Foretarsus slender; forebasitarsus 5x as long as broad and with 5 very short rake setae; midbasitarsus straight; T-VII with large, nearly triangular spiracular lobes which are narrowly separated medioventrally (fig. 3c), T-VII notched posteriorly; S-II with a well developed and sharp medioventral pair of projections; S-VI with posterior edge angled backward medially (fig. 3m), S-VIII with a strong medioventral tooth (fig. 3k); gonostylus moderately broad (fig. 3i), not hairy beneath.

Female. Length 9.5-12.0 mm. Markings about as in male; posterior projections of black bands on T-II-III narrow, on II (fig. 3n) sometimes reduced to dots; clypeus heavily silvered. Forebasitarsus 2.5x as long as broad, with 3 well developed preapical rake setae (fig. 3L); tarsomere V with numerous short erect setae beneath; T-VI unusually narrow posteriorly, punctation moderate and well spaced but thicker and coarser laterally.

Geographic range. Upper Sonoran and Transition life zones of California in San Diego, Imperial, and Riverside counties, north to Mono County in the east and Stanislaus County in the west (map 2). Out-of-state records are from Nevada.

Records. Holotype male, 2 mi. e. Cajon Junction, San Bernardino Co., California, VIII-11-56 (J. C. Hall, UCD). Paratypes, 42 males, 38 females. California: STANISLAUS CO.: Del Puerto Canyon, VII-24-75 (N. Smith, R. Kimsey); SAN BENITO CO.: Idria, VI-29-54 (C. MacNeill), VI-8-81 (N. Smith); FRESNO CO.: 18 mi. sw. Mendota, VII-19-75 (J. Powell); Mineral Springs (J. Halstead); 8 mi. sw. Coalinga; MONO CO.: Topaz Lake, VI-26-57 (J. MacSwain); Mammoth, Convict Lake, VIII-2-36 (R. Bohart); INYO CO.: Bishop, VI-24-49 (L. Isaak); Antelope Springs, VII-24-60 (P. Hurd), VII-1-61, on Eriogonum (J. Powell); Batchelder Springs, VI-16-60 (H. Court); Lee Flat, VIII-29-38 (A. Menke, L. Stange); 7 mi. n. Parchers Camp, VI-30-61 (J. Buckett, G. Stage); Big Pine, VI-18-29 (E. Van Duzee), VII-11-58, on Eriogonum (R. Bohart); Big Pine Creek, VII-10-61 on Eriogonum (R. Bohart), VII-15-77 (N. Smith); Lone Pine, VI-6-37 (W. Reeves); 3 mi. n. Lone Pine, VII-8-61 (D. Miller); Owens Valley, VII-2-36, on Eriogonum (R. Bohart); Little Lake, IX-3-56 (R. Bohart); KERN CO.: Kernville, VIII-19-52 (T. Haig); Cameron Road, VIII-3-82 (W. Tyson); SAN BERNARDINO CO.: Mojave Desert, X-6-35, on Penstemon (E. Linsley); 10 mi. w. Barstow; New York Mts., V-11-79 (Brown and Faulkner); Cajon Junction, VII-11-56 and VI-7-58 (E. Schlinger); Morongo Valley, VIII-3-33 (P. Timberlake), VII-3-65 (J. Bath); Yucca Valley, VII-3-33 (H. McKenzie); Deep Creek, VIII-1-36 and VIII-1-56 (E. Schlinger, J. Hall), VII-12-64 (R. Snelling); Black Canyon and Cedar Canyon, IX-4-80 (T. Griswold); Home Canyon, Oak Springs, Arrastre Canyon, and Horse Canyon, VIII-6-78 (T. Griswold); 5 mi. s. Cima, VI-10-80 (T. Griswold); Granite Mts., V-29-81 (T. Griswold); 3 mi. e. Yucaipa, VII-5-77 (T. Griswold); LOS ANGELES CO.: Tanbark Flat, VI-24-50 (A. McClay); RIVERSIDE CO.: Pinyon Flat, VII-13-40 (P. Timberlake), VII-23-62 (E. Montgomery); Snow Creek Road, V-26-76 (L. Bezark); Anza, VII-29-38 (D. Craik); SAN DIEGO CO.: Warner Springs, VII-8-56, on Eriogonum (R. Bohart); Jacumba, VIII-2-61, on Eriogonum (J. Gillaspy); Pine Valley; Mt. Jupiter, VIII-23-46 (N. Gratz); IMPERIAL CO.: In-Ko-Pa Gorge, VI-24-54 (W. McDonald). Nevada: WASHOE CO.: Pyramid Lake, VI-20-74 (B. Villegas); LINCOLN CO.: 6 mi. e. Panaca; 3 mi. s. Caliente, VII-11-78 (T. Griswold); CLARK CO.: 12 mi. s. Mesquite, VIII-25-59 (F. Parker).

Discussion: Glenostictia angulifera is related to angulata, as indicated by the angled S-VI in males of both species. Another similarity is the definitely yellow markings dorsally on the abdomen of males; this differentiates them from males of the common pulla, in which the yellow is infused with whitish. On the basis of markings, angulifera males resemble those of gilva, and attention must be given to the shape of S-VI as well as to the much broader spiracular lobes of angulifera (compare figs. 2k, c, and 3m, c). From angulata males, the all-or-nearly-all yellow clypeus, and mostly reddish rather than brown undersurface of the flagellum, are important differences. Females are easily separated from those of angulata by the much more heavily silvered clypeus of angulifera. The most similar forms are pulla and gilva. There are subtle differences in the clypeus, but the most useful characters in angulifera are the 3 versus 4-5 well developed preapical comb setae of the forebasitarsus (compare figs. 3L, and 10h), and the more restricted black area of T-II.

Glenostictia argentata (C. L. Fox)
(Fig. 4; map 3)

Stictiella argentata C. Fox 1923:434. Holotype male, "Angeles Bay, Lower California" (CAS).

Geographic range. California from Fresno County (near Coalinga) south on the west side of the Sierra Nevada, and from Inyo County (Bishop) south on the east side. The species occurs also in s. Utah, w. Nevada, w. Arizona, Baja California, and Sonora (map 3).

California records. FRESNO CO.: Jacalitos Canyon; KERN CO.: Buttonwillow, Inyokern; INYO CO.: Bishop, Big Pine, Lone Pine, Panamint Springs, Olancha; SAN BERNARDINO CO.: Kelso, Cronise Valley, Kramer Mts., Hesperia, Providence Mts., Colton Hills, 5 mi. s. Cima, Twentynine Palms, Victorville; LOS ANGELES CO.: Llano; RIVERSIDE CO.: Riverside, Eagle Mts., Palm Springs, Palm Desert, Blythe, Magnesia Canyon, Indio, Mecca; SAN DIEGO CO.: Borrego, Scissors Crossing, Jacumba; IMPERIAL CO.: Glamis, Westmorland, Gordons Well, Mt. Signal, Andrade.

Discussion. The dense silvering of the scutum which obscures punctation is unique in the pulla group. In both sexes, the pale abdominal markings are partly to almost wholly whitish. In males the spiracular lobes of T-VII in posterior view are thickened and wedgelike. After dissection, the posterior border of the lobe is seen to be emarginate

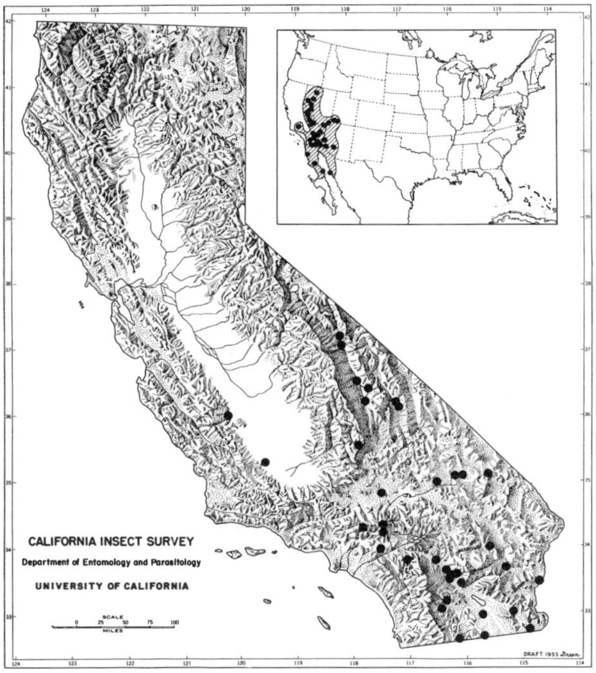

Map 3. California distribution of Glenostictia argentata (C. Fox). Inset: overall distribution.

Map 4. California distribution of Glenostictia arizonae Bohart. Inset: overall distribution.

(fig. 4c). The female forebasitarsus has 4 well developed preapical rake setae and a weak fifth one (as in fig. 10h).

Alcock (1975) found one burrow in a stabilized sand dune in Maricopa Co., Arizona. It was a sloping tunnel, 15 cm deep and ending in a single cell stocked with flies (primarily Bombyliidae). A temporary closure was made each time the female left the nest, and provisioning was progressive.

Flower records are Croton, Petalonyx, Eriogonum, and Tamarix. Captures have been made mostly near sand dunes and in each month from April to August. We have seen about 100 males and 65 females.

Glenostictia arizonae Bohart
(Fig. 5; map 4)

Glenostictia arizonae Bohart 1983:487. Holotype male, Rucker Canyon, Chiricahua Mts., Cochise Co., Arizona (UCD).

Geographic range. Upper Sonoran and Transition life zones in s. California, Arizona, w. New Mexico, and nw. Mexico (map 4).

California record. SAN DIEGO CO.: 1 male, Pine Valley, no date (W. J. Chamberlin, UCD).

Discussion. The pictifrons group, of which arizonae is a member, is especially notable for the modified foretarsi in the male. Foretarsals II-IV are somewhat flattened and expanded, as well as posteriorly blackened and often lobed (fig. 5g). Other characters are: clypeal margin nearly straight above medially (fig. 5d), midocellar plane about as broad as long, posteromedial pale scutal marks absent or rarely present and minute, hindfemur without significant erect hair beneath, male midbasitarsus strongly curved (fig. 5f), male T-VII with broad spiracular lobes (fig. 5c), male S-II edentate, male S-VIII with a medioventral tooth or spine (fig. 5m), female clypeus with a transverse yellow band above.

Glenostictia arizonae is closely related to pictifrons, and the two species may occasionally be taken together in the mountains of s. Arizona. Both have male foretarsals III-IV with small and pointed lateral lobes (fig. 5g). Also, the medioventral projection of S-VIII in the male is spinelike (fig. 5m). However, in arizonae males there are no black marks above on foretarsals I-II or on hindtarsals III-IV. Additionally, the clypeus of male arizonae is whitish rather than yellow. The female of arizonae has the mediodorsal pale spot on T-II-III separate or nearly so, and the black basal border of S-II is angled back medially (fig. 5h), but not so in *pictifrons*.

G. arizonae is the only member of the *pictifrons* group to be found in California, where it is known only from 1 male collected in the mountains of San Diego County. The species is moderately abundant in the mountains of s. Arizona. Bohart collected a long series in Rucker Canyon of the Chiricahua Mountains along with 1 male *pictifrons*. At a locality some 20 miles east, near Apache, 3 species of the group were flying together over patches of *Eriogonum deflexum*: *G. pictifrons*, *nigriloba*, and *vechti*.

Collections have been made from July to September, mostly in August. A variety of flowers were visited, including *Verbesina*, *Eriogonum*, *Encelia*, and *Melilotus*. Gillaspy et al. (1962) gave the prey of the closely related *pictifrons* as Bombyliidae (*Villa*), and Alcock (1975) reported "*pictifrons*" as preying on small Diptera. Nothing has been reported on prey or nesting habits for any other member of the group. We have seen a total of about 60 males and 100 females.

Glenostictia bifurcata (C. Fox)
(Fig. 6; map 5)

Stictiella bifurcata C. Fox 1923:431. Holotype male, "Puerto Refugio, Angel de La Guardia Island, Lower California" (CAS).
Stictiella bifurcata albicera C. Fox 1923:432. Holotype male "Guaymas, Sonora," Mexico (CAS).
Stictiella directa C. Fox 1923:433. Holotype male, "Angeles Bay, Lower California" (CAS).

Geographic range. In s. California, s. Nevada, and s. Arizona. Mexican states are Sonora, Baja California and Baja California Sur. Most localities, especially those in California, are in the Lower Sonoran Life Zone (map 5).

California records. SAN BERNARDINO CO.: Van Winkle Mts., Camp Baldy, Morongo Valley; RIVERSIDE CO.: Palm Springs, Tahquitz Canyon, Shavers Well, Carrizo Creek, near Palm Desert, Magnesia Canyon; SAN DIEGO CO.: Borrego Valley, Scissors Crossing, San Felipe Valley, Mt. Laguna, Ocotillo, Dulzura, Yaqui Well, Jacumba; IMPERIAL CO.: In-Ko-Pah Gorge, Black Mts., Chocolate Mts.

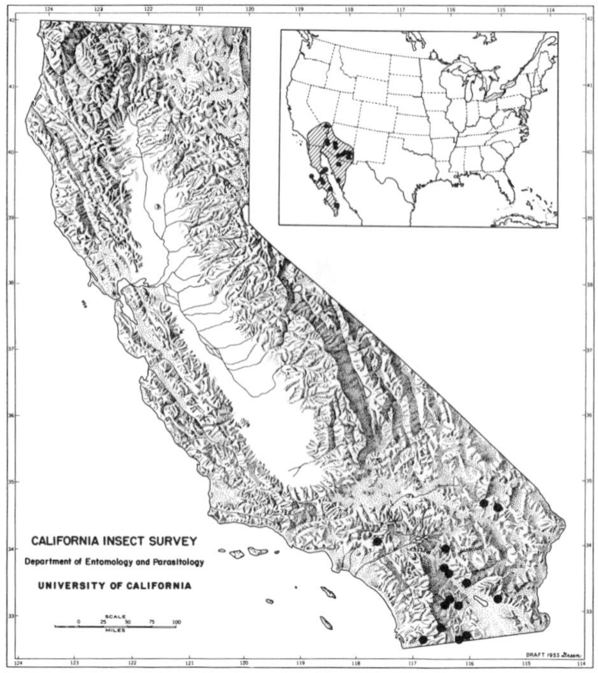

Map 5. California distribution of *Glenostictia bifurcata* (C. Fox). Inset: overall distribution.

Discussion. *Glenostictia bifurcata* is one of the more easily recognizable species in the genus. In both sexes, there is a pair of pale spots (sometimes coalesced) on T-I (fig. 6e). Also, the male hindfemur has abundant erect hair beneath (fig. 6o). In the female, the hair is present, but less obvious. Nearly all specimens have narrow discal spots or stripes on the scutum. The area around the tentorial pits is darkened, and in females it usually expands to a pair of dark spots at the base of the clypeus. Other distinctive features are, in the male: a median but distally bifurcate projection on S-II (fig. 6n), straight midbasitarsus (fig. 6f), rather narrow spiracular lobes of T-VI (fig. 6c), and a stout median tooth on S-VIII (fig. 6m).

Specimens have been collected at flowers of *Heliotropium*, *Prosopis*, *Petalonyx*, and *Tamarix*. Collections have been made in every month from April to August. We have seen 58 males and 70 females.

Glenostictia bituberculata (J. Parker)
(Fig. 7; map 6)

Stictiella bituberculata J. Parker 1917:36. Holotype male, San Bernardino Co., California (according to Cockerell 1918) (USNM).

Geographic range. California, Arizona, Baja California and Sonora.

California records. ALAMEDA CO.: Tesla; TUOLUMNE CO.: Hetch Hetchy Dam; FRESNO CO.: Coalinga, Jacalitos Canyon, Mt. Diablo Range, Little Panoche Creek; SAN BENITO CO.: near Panoche Pass; TULARE CO.: Goshen Junction; INYO CO.: Westgard Pass, Deep Springs Lake, Antelope Springs, Owens Valley, Darwin Falls; SAN BERNARDINO CO.: Van Winkle Mts., Colton, 10 mi. e. Twentynine Palms; RIVERSIDE CO.: Joshua Tree National Monument, Pinyon Flat, Santa Rosa Mt.; SAN DIEGO CO.: Mt. Laguna and 5 mi. n.

Discussion. *Glenostictia bituberculata* appears to be related to *terlinguae* and *bifurcata*, particularly the former. The range of *terlinguae* is Arizona to w. Texas, and both species occur in Arizona. The male of *terlinguae* has 1 sharp median projection (rather than 2 or none) on S-II, and there are free submedian yellow spots on T-I-III at least. Females of *terlinguae* have the lateral branches of the yellow discal U on the scutum bluntly rounded in front instead of tapering as in *bituberculata*. Since specimens of both species have been taken in the Hualpai Mts. of Mohave County, Arizona, it is possible that *terlinguae* may be found in mountainous areas of San Bernardino or Inyo counties of California. In *bifurcata*, both sexes have the hindfemur more hairy than in *bituberculata*, and the female clypeus is ordinarily black-spotted. The pair of projections on male S-II, from which the name is derived, vary in degree of prominence. In the 20 males examined, 4 are without tubercles. *G. bituberculata* is a relatively large species, with body length 15-20 mm. Collections have been made in each month from May to September. Flowering plants visited were *Chrysothamnus*, *Lepidospartum*, and *Eriogonum*. We have studied about 20 males and 30 females.

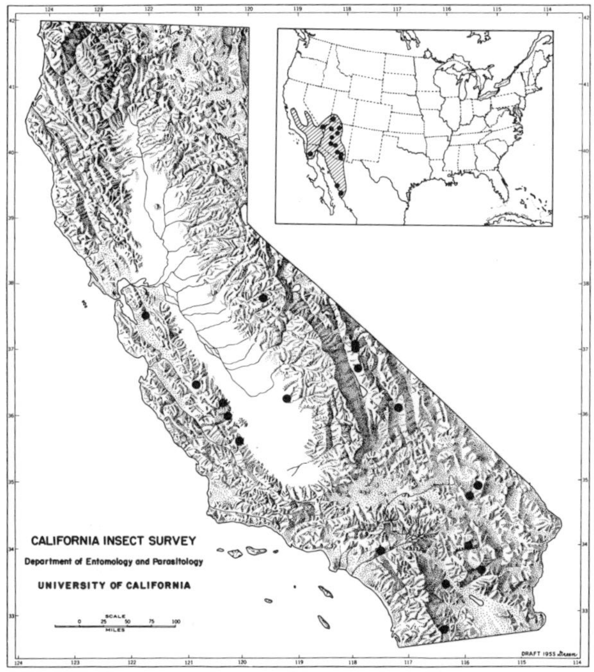

Map 6. California distribution of *Glenostictia bituberculata* (J. Parker). Inset: overall distribution.

Glenostictia californica Bohart
(Fig. 8; map 7)

Glenostictia californica Bohart 1983:485. Holotype male, 14 mi. s. Baker, San Bernardino Co., California (UCD).

Geographic range. Upper Sonoran localities in se. California, se. Nevada, and sw. Arizona (map 7).

California records. 12 males, 12 females. INYO CO.: 5 mi. nw. and 8 mi. se. Panamint Springs, V-8-58 (R. Bohart), V-15-69 (P. Opler, P. Rude); Darwin and 6 mi. nne., V-12-69 (J. Powell, P. Rude); 2 mi. ne. Lone Pine; Surprise Canyon, V-9-58 (R. Bohart); Death Valley National Monument, IV-28-80 (L. and R. Kimsey); SAN BERNARDINO CO.: 4 mi. s., 14 mi. s., and 15 mi. se. Baker, IV-16 to IV-23-69 (M. and J. Wasbauer, and E. Schlinger); 7 mi. sw. Kelso (M. and J. Wasbauer); 10 mi. ne. Earp, IV-18-64 (R. L. Langston).

Discussion. *Glenostictia californica* is a rather small species which rarely exceeds a length of 11 mm in the male and 10 mm in the female. Other species characteristics are as follows: clypeal margin nearly straight above

Map 7. California distribution of *Glenostictia californica* Bohart. Inset: overall distribution.

Map 8. California distribution of *Glenostictia clypeata* Gillaspy. Inset: overall distribution. Extreme southern records from Colima and Guerrero are not shown.

medially (fig. 8d), midocellar plane about as broad as long, no posteromedian pale scutal marks, hindfemur without significant erect hair beneath, clypeus all white in both sexes, galea longer than eye height, male midbasitarsus straight (fig. 8f), male T-VII spiracular lobes rather broad (fig. 8c), male S-II edentate, male S-VIII with a medioventral sharp projection (fig. 8m).

Although the characters of *G. parva* agree with most of the above, a close relationship is doubtful. Both sexes of *parva* have the galea shorter than eye height. Furthermore, males of *parva* have the midbasitarsus strongly curved rather than straight.

Nothing is known of the biology of this species. Specimens have been collected in April and May (*parva* flies in August and September). We have examined 12 males and 15 females.

Glenostictia clypeata (Gillaspy)
(Fig. 9; map 8)

Stictiella clypeata Gillaspy 1959:187. Holotype male, 33 mi. e. Deming, Dona Ana Co., New Mexico (CAS).

Geographic range. Lower and Upper Sonoran life zones in se. California, Nevada, Arizona, New Mexico, Texas, and n. Mexico.

California records. 27 males, 6 females. SAN BERNARDINO CO.: Needles, VII-18-77 (R. Schuster); RIVERSIDE CO.: 4 mi. e. Mecca, VII-?-34 (G. Bohart); Ripley, VIII-19-46 (P. Hurd); IMPERIAL CO.: Meloland (Experimental Farm), V-VI, 1911-12 (J. Bridwell); Laguna Lake, VI-11-50; Holtville, VIII-11-17; Calexico, VIII.

Discussion: *Glenostictia clypeata* and *tenuicornis* constitute the *tenuicornis* group, in which the midocellar plane is longer than broad (fig. 9b), and the clypeus is slightly peaked above (fig. 9i) and in the female is flattened on either side of the middle. Male tarsi are unmodified and male S-II is not dentate. Both species have an abundance of yellow markings, but in *clypeata* the tergal markings are often arranged in a somewhat checkered pattern (fig. 9e). Male *clypeata* are easily distinguished by the midfemur, which is fairly smooth beneath for its whole length, rather than ridged and distally notched. In addition, S-VIII has only a carina midventrally in *clypeata* (fig. 9m), but a sharp spine in *tenuicornis*. Females are much more difficult to separate. In addition to the usually more extensive black markings of the terga with consequent "checkering," *clypeata* has the sublateral reticulation of T-II (seen in side view) more fine than that on T-III. In *tenuicornis*, the reticulation is about equally coarse on both terga. Both species have the clypeus with extensive silvery micropubescence, especially in the flattened areas.

As a possible extension of the geographical range given above, we have seen a series of specimens from southern Mexico: Guerrero (Mexcala) and Colima (Colima). These have the checkered pattern common to *clypeata* but they are much darker than usual and may represent a distinct subspecies.

Prey has been reported by Gillaspy et al. (1962) to be Diptera in the families Apioceratidae and Syrphidae (*Eristalis*). The flight period is July to September, but there are a few records for May and June in Imperial County, California. We have seen about 150 males and 155 females of this relatively abundant species.

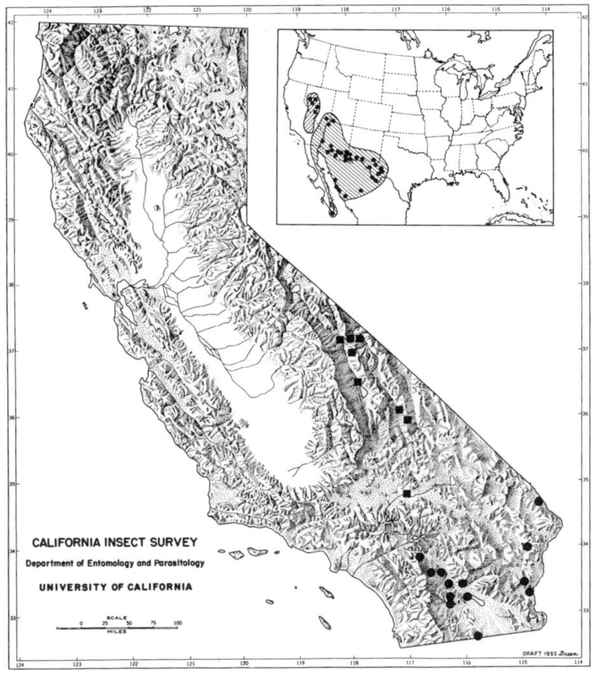

Map 9. California distribution of *Glenostictia gilva gilva* Gillaspy (black circles) and *G. gilva albescens* Bohart (black squares). Inset: overall distribution.

Glenostictia gilva gilva Gillaspy, new status
(Fig. 10; map 9)

Glenostictia gilva Gillaspy 1963a:198. Holotype male, Congress Junction, Arizona (KU).

Geographic range. California in the Coachella, Borrego, and Imperial valleys, and along the far se. border; also in s. Nevada, sw. Utah, and s. Arizona to w. Texas. In Mexico it is known from Baja California Sur, Sonora, Sinaloa, and Chihuahua. Most records are from desert localities in the Upper Sonoran and Lower Sonoran life zones.

California records. Representative localities are: SAN BERNARDINO CO.: mouth of Deep Creek; RIVERSIDE CO.: Andreas Canyon, Magnesia Canyon, Deep Canyon (near Palm Desert), Desert Center; SAN DIEGO CO.: The Narrows, Borrego Valley; IMPERIAL CO.: Palo Verde.

Discussion. Differences from subspecies *albescens* and from *pulla* are given under their discussions. Coloration and extent of markings on the antenna, clypeus, scutum, legs, and abdominal dorsum (fig. 10a, d, e, n, o) in one or both sexes are critical in separating the forms of this group.

Habits of *Glenostictia gilva* have been given by Evans (1966) and Alcock (1975). Burrows may be 11 cm long in powdery soil (Evans 1966), or 24-30 cm in stabilized sand (Alcock 1975). The tumulus is scattered after the burrow is constructed. The single cell is stocked with Bombyliidae (*Aphoebantus*) (Evans 1966) or "mostly Bombyliidae" (Alcock 1975). Provisioning is progressive.

Flowers visited include *Eriogonum* and *Tamarix*, and the flight period is April to October. We have seen about 50 males and 40 females of this subspecies.

Glenostictia gilva albescens Bohart, new subspecies
(Fig. 10, except n, o; map 9)

Male holotype. Agreeing structurally with males of *gilva* sensu stricto, including rounded posterior edge of S-VI (fig. 10p) and rather narrow spiracular lobes of T-VII. Coloration differences are: no discal pale spots on scutum (present but small in a few paratypes), whitish bands of terga extending to lateral edge.

Female. Agreeing with females of *gilva* sensu stricto, including rake setae of forebasitarsus (fig. 10h), but terga extensively whitish with only tinges of yellow.

Geographic range. Se. California from Riverside County, where an overlap with other subspecies occurs, north through the Mojave Desert and into the Owens Valley and adjacent areas to the east (map 9). It occurs also in w. Nevada.

Records. Holotype male, 3 mi. e. Big Pine, Inyo Co., California, VII-10-53 (W. D. McClellan, UCD). Paratypes, 26 males, 15 females. California: MONO CO.: Mammoth, VII-?-33 (G. and R. Bohart); INYO CO.: Mazourka Canyon in Inyo Mts., VII-2-53 (J. MacSwain, W. McClellan); Darwin Falls, VII-15-53 (E. Schlinger), VII-12-57 (E. Schlinger), V-17-70 (E. Grissell); 3 mi. n. Big Pine, VII-10-53 (W. McClellan), VII-13-61 (D. Miller); near Antelope Springs, VI-15-61 (C. Toschi), VII-11-67 (R. Bohart), no date (G. Stage); 3 mi. n. Lone Pine, VII-6-61 (J. Powell, D. Miller); Lone Pine, VI-1-37 (N. Frazier); 4 mi. e. Independence, V-30-77 (E. and J. Linsley); 15 mi. s. Panamint Springs, IV-28-53, on *Petalonyx* (P. Hurd); Owens Valley, VIII-2-36 (G. and R. Bohart); "Inyo Co.," VII-29-62 (T. Talbot); SAN BERNARDINO CO.: Cronise Valley, V-27-50; Barstow, IX-14-35 (P. Timberlake); RIVERSIDE CO.: Whitewater, VII-9-50 (P. Hurd); Mecca. Other paratypes, 22 males, 4 females. Nevada: WASHOE CO.: Nixon, VI-3-62 (R. Bohart, R. Gill, M. Irwin); 3 mi. n. Nixon, VI-25-60 (A. Menke, F. Parker); Pyramid Lake, VI-25-60, on *Cleome* (A. Menke), VI-25-75 (B. Villegas); Wadsworth, VI-25-77 (R. Bohart); CHURCHILL CO.: 5 mi. w. Hazen, on *Tamarix* (R. Bohart); 23 mi. e. Fallon, VII-21-58 (E. Linsley); 25 mi. e. Fallon, VI-19-80 (L. Hanks); ESMERALDA CO.: 7 mi. n. Dyar, VII-2-58 (R. Bechtel); 4 mi. n. Dyar, VII-8-65 (R. Bechtel).

Discussion. In *albescens*, the abdomen is infused with whitish in both sexes instead of yellow as in *gilva gilva*. This feature seems constant in our material.

Glenostictia megacera (J. Parker)
(Fig. 11; map 10)

Stictiella megacera J. Parker 1917:49. Holotype male, "Arboles, Colorado" (USNM).

Geographic range. Western United States as far east as Colorado, mostly in Upper Sonoran

Map 10. California distribution of *Glenostictia megacera* (J. Parker). Inset: overall distribution.

and Transition life zones. Mexican records are 4 mi. w. Cercado in Nueva Leon and Tepoca Bay in Sonora.

California records. MODOC CO.: Alturas; MONO CO.: Topaz Lake; INYO CO.: Bishop, Big Pine, Deep Springs, Antelope Springs, Batchelder Springs, Lone Pine; SAN BERNARDINO CO.: Windmill Station, Providence Mts., Van Winkle Mts., New York Mts., Cedar Canyon, Baldwin Lake, 25 mi. w. Needles, Baldy Mesa, Victorville, Adelanto, Ord Mts., Morongo Valley, Twentynine Palms; LOS ANGELES CO.: Llano; RIVERSIDE CO.: Pinyon Flat.

Discussion. *Glenostictia megacera* has many distinctive characters, particularly in the male, and there doesn't seem to be any close relative. In both sexes, the clypeus is peaked above to a point level with the lower edge of the antennal sockets (fig. 11d). Also, the midocellar plane is moderately long oval (fig. 11b), the LID is a little greater than the eye breadth, and the galea is about as long as the eye height. Body length is moderate, 13-16 mm in males, 11-14 in females. Males may be characterized as follows: F-I distad, F-II, and F-III basad distinctly broadened (fig. 11a); clypeus and labrum yellow; scutum dark except over tegulae; midbasitarsus and hindbasitarsus not curved but unusually broadened (fig. 11f); hindfemur with abundant, short, erect hair beneath; T-VII spiracular lobes broad (fig. 11c), pale basally; S-I-VI with abundant short, erect pubescence; S-II with a strong, stout medial spine (fig. 11i); S-VIII quite narrow, middle prong nearly straight, lateral projections short (fig. 11m, n), medioventral area with a raised carina but no tooth. Except for the clypeus, females are not so unusual: clypeus yellow above, becoming whitish below, slightly flattened on either side of middle where moderate punctures bear short, erect hair which appears woolly in lateral view; forebasitarsus about 2.7x as long as broad, 5 well developed preapical rake setae; scutum with a pair of small to medium anterodiscal yellow spots, but no posteromedial ones; T-II with yellow band broadly complete, black area ending in a rounded or weakly pointed lobe (fig. 11e).

There are some similarities to bifurcata, which has a stout median (but bifurcate) projection of the male S-II, and the female clypeus with erect pubescence. However, differences are many. The thickened F-I-III of male megacera (fig. 11a), and the rounded rather than tripronged black mark of T-I in the female (fig. 11h), are immediate points of separation.

G. megacera is a widespread species in the western United States. It flies from April to August, but most captures have been in June and July. Prey is unknown; flower records include Eriogonum, Achillea, Melilotus, Senecio, Ribes, Cleome, Thelypodium, and Chamaebatieria. We have seen about 40 males and 50 females.

Glenostictia pulla (Handlirsch)
(Fig. 12; map 11)

Monedula pulla (Handlirsch) 1890:149. Holotype female, "Californien," Hamburg Museum (destroyed). Neotype female, Arroyo Seco Camp, Monterey Co., California, V-27-56 (R. M. Bohart, UCD). Here designated to avoid possible confusion with G. angulata, angulifera, and gilva.

Monedula usitata W. Fox 1895:371. Lectotype male (Cresson, Jr. 1928), "Nevada" (ANSP).

Geographic range. Widespread in California, being replaced and slightly overlapped by the closely related gilva gilva in the extreme southeast, and by gilva albescens in

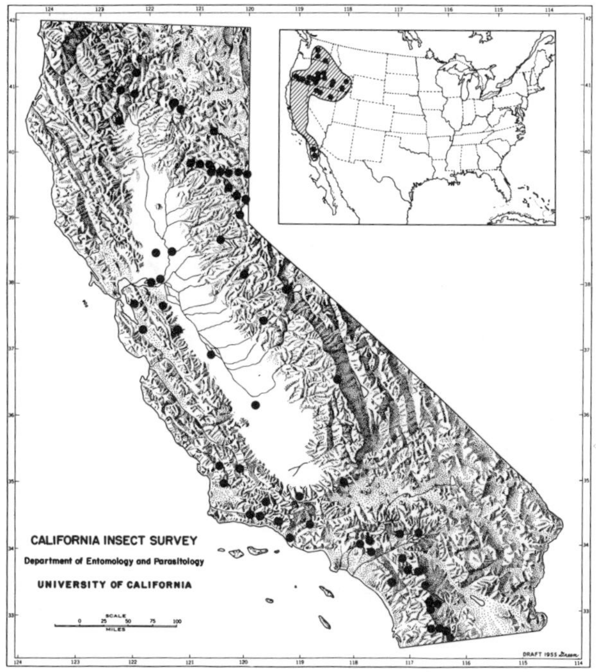

Map 11. California distribution of Glenostictia pulla (Handlirsch). Inset: overall distribution.

the Mojave Valley and Owens Valley. G. pulla ranges east into n. Nevada, then north into Idaho, Oregon, and Washington. It has been collected also in the Sierra San Pedro Martir of Baja California. It occurs mostly in the Upper Sonoran to Canadian life zones.

California records. Representative localities are: MODOC CO.: Wrightwood; SISKIYOU CO.: Willow Creek Mt.; SHASTA CO.: Redding; TRINITY CO.: Eagle Creek; LASSEN CO.: Bridge Creek Camp; TEHAMA CO.: Manton; PLUMAS CO.: Quincy; SIERRA CO.: Sierraville; NEVADA CO.: Sagehen Creek; EL DORADO CO.: Ice House Road; ALPINE CO.: Wolf Creek; TUOLUMNE CO.: Mather; MONO CO.: 11 mi. n. Bridgeport; ALAMEDA CO.: Corral Hollow; STANISLAUS CO.: Del Puerto Canyon; INYO CO.: Lone Pine Creek at 8,500-ft. elev.; MONTEREY CO.: Arroyo Seco Camp; MARIPOSA CO.: Yosemite Valley; SAN BENITO CO.: 13 mi. sw. Idria; SAN LUIS OBISPO CO.: Creston; SANTA BARBARA CO.: Los Prietos; LOS ANGELES CO.: Tanbark Flat; RIVERSIDE CO.: Idyllwild; SAN DIEGO CO.: Jacumba.

Discussion: Within its western range, pulla is one of the most abundant species of Glenostictia. It is the nominal species of the pulla group, and as such has the clypeus peaked and extending upward between the antennal sockets (fig. 12d). In addition, the hindfemur has almost no erect hair beneath, and the midocellar plane is long oval (fig. 12b). The female clypeus is somewhat flattened on either side of the middle, and foretarsal V is bristly beneath. The male has the tarsi simple, with the midbasitarsus straight. S-II in the male is bidentate (fig. 12j).

As a species, pulla is distinguished in the male from some other members of the group (angulata, angulifera, argentata) by the much narrower ventral spiracular lobes of T-VII (compare figs. 11c, and 3c), from the first two by the rounded posterior margin of S-VI; and from argentata by the punctate or heavily reticulate scutal sculpture which is not obscured by pubescence.

The female of pulla has the clypeus rather densely silvered, contrary to angulata; T-II with broad black projections from the basal black band, contrary to angulifera; and scutal differences from argentata, as in the male.

G. pulla and gilva seem most closely related since they have the same narrow spiracular lobes and blunt S-VI in the male. Consequently, separation depends on color and extent of markings. In the male and occasionally the female of pulla, the clypeus has extensive dark markings (fig. 12d) or may be nearly all dark. The femora of males are black-streaked anteroventrally (often in females, also). Further, the male terga are whitish, toward the center at least. In gilva, the clypeus is all pale or with small lateral black areas, and the femora are unmarked anteroventrally. In addition, the terga of both sexes are yellow in gilva gilva, although whitish in gilva albescens. The female of the latter has the scutum with small pale discal spots. Females of pulla and gilva gilva are less easily separated, but the yellow discal spots of the scutum average considerably larger in the latter (figs. 10o, 12o). One additional but not entirely consistent species difference is the tendency of pulla males to have the S-II projections more weakly developed than in gilva (figs. 10q, 12j).

The extent of black on the male clypeus of G. pulla is variable, but nearly always covers at least a third of the surface. In 50 specimens taken at random but each from a separate locality, 34% had the clypeus all black except for the narrow apical rim, and 66% had the black mark considerably excised anteromedially. Those with the darker clypeus generally came from more northern localities or from higher elevations.

The larva of pulla was described by Evans and Lin (1956). Habits of pulla have been reported by LaRivers (1942), Gillaspy et al. (1962), and Evans (1966). LaRivers made his observations in Nevada, Gillaspy in California and Idaho, and Evans in Utah. Nests are made in bare areas where the soil is a light powdery sand or sandy loam. Adults sleeping in clusters containing both sexes were seen by Gillaspy. Burrows are short (ca. 11 cm long), with the cell about 6 cm below the surface (Evans 1966). Provisioning is progressive, and a variety of flies are taken: Syrphidae (Eupeodes, Helophilus), Bombyliidae (Aphoebantus, Geron, Lepidanthrax), Sarcophagidae (Senotainia, Sarcophaga), Tachinidae (Stomatomyia), Muscidae (Hylemyia), Therevidae (Psilocephala). The female wasp scatters the tumulus and closes the nest temporarily while hunting for prey. The egg is laid on the side of the thorax of the first fly in the cell. A variety of flowers are visited, but particularly Eriogonum. The flight period is May to September. We have seen about 1,100 males and 600 females of this relatively abundant form.

Glenostictia scitula (W. Fox)
(Fig. 13; map 12)

Monedula scitula W. Fox 1895:369. Lectotype female, Tucson, Arizona (USNM). Here designated by R. M. Bohart.
Monedula villosa W. Fox 1895:37. Holotype male, "Lower Purissima, Lower California" (ANSP).

Geographic range. S. California to w. Texas, south into n. Mexico and north into Nevada and s. Utah. Most localities are in the Lower Sonoran Life Zone.

California records. INYO CO.: Darwin Falls, Surprise Canyon, Warm Sulfur Springs, Death Valley (Midway Well, Stovepipe Wells, Furnace Creek); SAN BERNARDINO CO.: 15 mi. w. Baker, Cronise Valley, Afton Canyon, Manix, Barstow, Renoville, Needles, Parker Dam, Twentynine Palms; RIVERSIDE CO.: Thousand Palms, Palm Desert, La Quinta, Mecca, 6 mi. w. Indio, Coachella, Thermal, Shavers Well, Hopkins Well, 6 and 18 mi. w. Blythe, Ripley; SAN DIEGO CO.: Borrego Valley, Yaqui Well, Jacumba; IMPERIAL CO.: Palo Verde, Kane Springs, Westmorland, 4 mi. nw. Imperial, Holtville, 12 mi e. Heber, near Bard.

Map 12. California distribution of *Glenostictia scitula* (W. Fox). Inset: overall distribution.

Discussion. This is one of the more common species of *Glenostictia* in southwestern desert regions. It is a relatively small species, with males averaging 11 mm in length and females 9.5 mm. There is considerable disparity in markings of the sexes. Females are predominantly yellow, particularly on the abdomen (fig. 13i), and there is a well developed (sometimes broken) discal yellow U on the thorax (fig. 13n). Males have the thorax much darker, but usually with small discal spots on the scutum. Furthermore, the male abdomen is characteristically whitish on the dorsum, grading laterally into yellow. Both sexes, but especially the male, have considerable erect hair on the clypeus (fig. 13j) and beneath the hindfemur (fig. 13h). Also, the clypeus angles up above the lower margin of the antennal sockets (fig. 13d). Other features in the male are the straight midbasitarsus (fig. 13f), weak to moderate pair of tubercles on S-II, narrow spiracular lobes of T-VII (fig. 13c), and well developed median tooth on S-VIII. Specimens are attracted to a variety of flowers such as *Larrea*, *Acacia*, *Sapindus*, *Parkinsonia*, *Baccharis*, *Lepidium*, *Heliotropium*, *Pluchea*, *Melilotus*, *Eriogonum*, *Medicago*, *Prosopis*, *Sisymbrium*, and *Condalia*.

G. scitula is primarily vernal. Collection records extend from March until June with an occasional later date; the species is most abundant in April and May. We have studied about 150 males and 70 females.

Gillaspy (in Gillaspy et al. 1962) and Evans (1966) reported on nesting habits, prey, and parasites in w. Texas. Evans (1964) described the larva. Nests were made in sandy loam or even in stony soil. There were 5 or more cells per nest, and as many as 40 prey per cell were provisioned progressively. Prey were especially bees of the genus *Perdita*, but other small Hymenoptera, Diptera, and even Cicadellidae were used also. Parasites were a chrysidid, *Parnopes concinnus* Viereck, and a miltogrammine sarcophagid fly.

Glenostictia tenuicornis (W. Fox)
(Fig. 14; map 13)

Monedula tenuicornis W. Fox 1895:368. Lectotype female (Gillaspy 1959:191), San Bernardino Co., California (USNM).

Geographic range. California, Nevada, Arizona, Baja California, and Baja California Sur, mostly in the Upper Sonoran Life Zone.

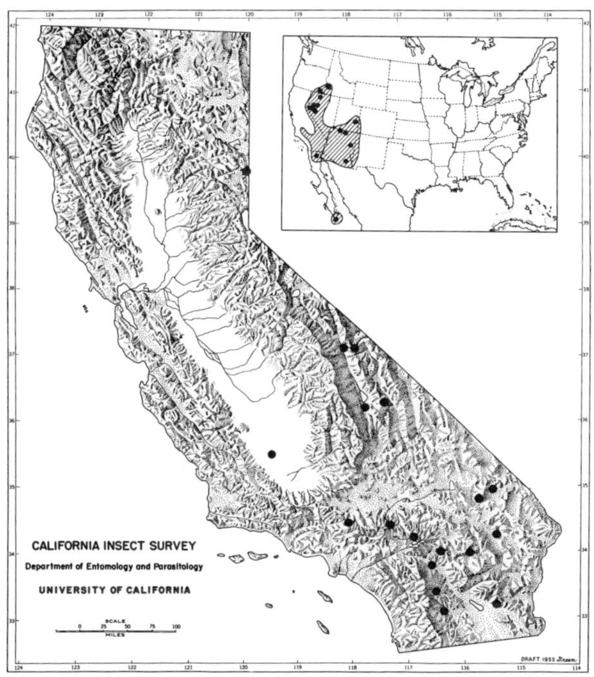

Map 13. California distribution of *Glenostictia tenuicornis* (W. Fox). Inset: overall distribution.

California records. LASSEN CO.: Constantia; INYO CO.: Big Pine, Deep Springs, Antelope Springs, 3 mi. w. Coso Junction; KERN CO.: 15 mi. nw. Shafter; LOS ANGELES CO.: Palmdale; SAN BERNARDINO CO.: 25 mi. s. Ivanpah, Victorville, Chubbock, Cushenbury Springs, Yucca Valley; RIVERSIDE CO.: Whitewater Canyon, Santa Rosa Mt.; SAN DIEGO CO.: Scissors Crossing; IMPERIAL CO.: Chocolate Mts.

Discussion. The characteristics of Glenostictia tenuicornis, and its close relationship to clypeata, have been discussed by Gillaspy (1959:191). Although tenuicornis is slightly yellower overall than clypeata, the two species are quite similar and together form the tenuicornis group. As indicated under clypeata, the notched male midfemur (fig. 14o), and presence of a sharp medioventral projection on the male S-VIII, are distinctive for tenuicornis. Females have the reticulation of T-II-III rather equally coarse as viewed laterally.

Little is known of the biology. Adults visit flowers of Eriogonum, Chrysothamnus, Heliotropium, Petalonyx, Lepidospartum, and Croton. They have been collected in each month from June to October. We have seen about 50 males and 30 females.

Genus Stictiella J. Parker

Monedula, Stictia of authors, in part.
Stictiella J. Parker 1917:21. Generotype: Monedula formosa Cresson, original designation.

The genus consists of 12 nearctic species, all but 2 of which (evansi, gillaspyi) are found in America north of Mexico. The California fauna includes 9 species.

Biology. In the 6 species for which biology is known, adult Lepidoptera are the prey, 2 species apparently preferring butterflies and 4 recorded as taking only moths. Publications of special value are Gillaspy et al. (1962), Evans (1964: larval descriptions, and 1966), Krombein (1966), and Alcock and Gamboa (1975).

Stictiella females nest in flat or slightly sloping dry, sandy soil. They occur in small aggregations, and females do not seem to spend the night in their burrows. One species has been reported to form clusters on vegetation, with both sexes participating. Leveling of the tumulus occurs in some species, and temporary closures are in place during prey gathering. One to two cells per nest seems to be the rule, but as many as 17 have been reported for S. formosa (Gillaspy et al. 1962). Recorded prey per cell have been 7 to 21, and mass provisioning appears to be the rule. However, delayed provisioning may occur when the egg, laid on the first prey, hatches prematurely. Natural enemies are miltogrammine Sarcophagidae (Senotainia) and sometimes mites.

Taxonomy. The important morphological characters of Stictiella are: (1) midocellar plane as broad as long or a little broader; (2) labrum nearly flat in profile; (3) palpal formula 6-4; (4) clypeus rather strongly convex in lateral profile; (5) arolium absent, or present but small (in female); (6) clypeal margin straight or concave medially above; (7) male midfemur carinate, and serrate or irregular beneath (fig. 17i); (8) male S-VIII with or without a midventral tooth or spine; (9) midbasitarsus distinctly curved (except evansi).

The somewhat narrower midocellar plane separates this genus from Microstictia. Males of Stictiella differ by their midfemur (character 7 above) from Glenostictia. Females are less easily separated, but a combination of characters 1, 2, 4, 5, and 6 above will distinguish them from other stictiellin genera.

A division of Stictiella into species groups can be made on the basis of various male characteristics and the presence or absence of an arolium in both sexes. The reduction of this adhesive pad in females may have been influenced during evolution by the type of prey (adult Lepidoptera) in which the scaled wings made adherence of the arolium ineffective.

1. The emarginata group. Arolium present but small in female. Male: foreleg distitarsus not expanded, and claws equal; S-II bidentate, S-VI not angled posteromedially, S-VIII dentate or spinose medioventrally. Included species: emarginata, villegasi.
2. The pulchella group. Arolium undeveloped. Male: foreleg distitarsus not expanded (except tuberculata), and claws equal;

S-II dentition various, S-VI angled back posteriorly and pointed (except _corniculata_), S-VIII not dentate medioventrally. Included species: _boharti_, _callista_, _corniculata_, _evansi_, _pulchella_, _tuberculata_.

3. The _flavescens_ group. Arolium present but small in female. Male: foreleg distitarsus not expanded, and claws equal; S-II not dentate, S-VI rounded or slightly emarginate posteromedially, S-VIII not dentate medioventrally. Included species: _flavescens_.
4. The _speciosa_ group. Arolium present but small in female, forewing a little longer than abdomen. Male: foreleg distitarsus expanded and claws equal, S-II bidentate, S-VI rounded posteromedially, S-VIII not or only weakly dentate medioventrally. Included species: _fergusoni_, _speciosa_.
5. The _formosa_ group. Arolium present but small in female, forewing hardly as long as abdomen. Male: foreleg distitarsus expanded, and claws unequal; S-II not dentate, S-VI a little emarginate posteromedially, S-VIII not dentate medioventrally. Included species: _formosa_, _gillaspyi_.

KEY TO SPECIES OF _STICTIELLA_

1. Males; 7 visible terga, 11 flagellomeres. 2

   Females, 6 visible terga, 10 flagellomeres 14

2. Tarsi with a distinct but sometimes small arolium (figs. 18f, 19f), S-VIII with medioventral angle or projection in some species. 3

   Tarsi with arolium undeveloped (fig. 16g), S-VIII without medioventral angle or projection (fig. 15m). 9

3. Foretarsal V not conspicuously flattened, not much broader than forebasitarsus; S-VIII various. 4

   Foretarsal V conspicuously flattened, (fig. 19g), much broader than forebasitarsus; S-VIII with a medioventral ridge which may project slightly at apex (fig. 19m) 6

4. Frons more than one-third of head breadth (fig. 18d), scutum without large discal yellow U, hindfemur with considerable and usually prominent erect hair posteroventrally (fig. 18h), S-VIII with well developed basomedian projection (U.S., Canada, Baja California). . . . . _emarginata_ (Cresson)

   Frons no more than one-third of head breadth (fig. 20d), scutum with large discal yellow U, hindfemur without noticeable erect hair posteroventrally, S-VIII various. 5

5. Labrum longer than F-I-II together, clypeus with microscopic pubescence only, S-VIII not dentate basomedially (fig. 20m), S-II simple (s. California, s. Arizona, Sonora, Baja California Sur) . . . ._flavescens_ Gillaspy

   Labrum a little shorter than F-I-II together, clypeus with some erect or semierect hair above, S-VIII strongly dentate basomedially (fig. 23m), S-II bituberculate (se. California) . . . . _villegasi_ Bohart

6. Foreleg claws similar, midtibia at widest point narrower than hindtibia, T-VII lateral edge evenly convex in dorsal view. 7

   Foreleg claws dissimilar, midtibia at widest point at least a little broader than hindtibia, T-VII lateral edge markedly concave in dorsal view. . . . 8

7. Forefemur with abundant, fine, erect hair on its undersurface, foretarsal V broadened and all black (Canada, central U.S. to Arizona) . . . . _speciosa_ (Cresson)

   Forefemur fringed but without conspicuous pubescence on its undersurface, foretarsal V broadened but medially translucent (fig. 19g) (w. U.S., nw. Mexico) . . . . _fergusoni_ Bohart

8. Midtibia broadest at distal fifth, foretarsomere I narrowed distally, foretarsomere V all dark (Nebraska to Arizona, n. Mexico). . . . . . ._formosa_ (Cresson)

   Midtibia broadest at distal third, foretarsomere I broadened distally, V black only toward apex (s. Arizona, Sonora, Sinaloa). . . . . gillaspyi Bohart

9. Midbasitarsus nearly straight (Sinaloa, Nayarit, Guerrero). . . evansi Gillaspy

Midbasitarsus curved and with a clump of strong bristles near base of curve (fig. 20f). 10

10. S-II with a median tubercle or projection 11

S-II simple, or with two submedian humps or projections. 12

11. S-II projection rounded at apex and hirsute (fig. 22i); foretarsal V expanded, dark, and flattened (fig. 22g); T-VII contorted, slightly incurved laterally in dorsal view (British Columbia, w. U.S., Baja California). . . . . tuberculata (W. Fox)

S-II projection micropubescent and sharp or merely a hump, foretarsal V yellow and not expanded (fig. 16g), T-VII evenly convex laterally in dorsal view (sw. U.S., n. Chihuahua)..callista J. Parker

12. Hindfemur concave ventrally (fig. 17h); S-VIII lateral projections stout, only a little acute (fig. 17n) (w. U.S. to e. California. . . . .corniculata Mickel

Hindfemur rather evenly convex ventrally; S-VIII with lateral projections slender and pronglike (fig. 21m). 13

13. F-X with lateral projection (fig. 15a), hindtibia with tuft of hair (fig. 15o) (s. California). . . . .boharti Gillaspy

F-X without lateral projection, hindtibia without tuft of hair distad (w. U.S., n. Mexico). . . . . .pulchella (Cresson)

14. Arolium small but distinct. 15

Arolium undeveloped. 21

15. Labrum a little shorter than F-I-II together, submarginal cell II with only a few scattered microsetae, scutum with a well developed yellow U (fig. 23h) (se. California). . . . . . .villegasi Bohart

Labrum a little longer than F-I-II together, second submarginal cell with many microsetae, scutal yellow markings various. 16

16. Hindfemur posteroventrally with some erect or semierect hair, LID about 1.5x eye breadth, scutum usually without strong yellow U (U.S., Canada, Baja California). . . . emarginata (Cresson)

Hindfemur with only a few short erect hairs, LID less than 1.5x eye breadth, scutum nearly always with well developed yellow U (fig. 20h) 17

17. Body nearly all yellow ventrally, clypeus with much smooth surface (s. California, s. Arizona, Sonora, Baja California Sur) . . . . . . flavescens Gillaspy

Body with considerable black ventrally, clypeus with obvious fine punctation and usually with scattered coarse punctures below middle. 18

18. Forewing a little longer than abdomen, F-II about 2.5x as long as broad. . . 19

Forewing hardly as long as abdomen, F-II about 2.2x as long as broad. 20

19. Hindtarsal V all or nearly all dark, pale dorsal abdominal markings nearly always infused with whitish (Canada, central U.S. to Arizona). . . speciosa (Cresson)

Hindtarsal V all or nearly all yellow, pale dorsal abdominal markings yellow (w. U.S., nw. Mexico)..fergusoni Bohart

20. Flagellum dull or brown toward apex beneath, galea about half yellow and half light reddish-brown, body length about 19 mm (Nebraska to Arizona, n. Mexico) . . . . .formosa (Cresson)

Flagellum broadly pale toward apex beneath, galea mostly dark brown, body length about 14 mm (s. Arizona, Sonora, Sinaloa). . . . . gillaspyi Bohart

21. Midtibia with dark spot at femorotibial joint (fig. 15h). 22

Midtibia yellow at femorotibial joint..23

22. S-II nearly all yellow, and clypeus all yellow; hindfemur flattened or a little

incurved on basal half beneath, ventral outline angled out (fig. 15i) (s. California. . . . . boharti Gillaspy

S-II about half black; clypeus with a pair of black basal spots; hindfemur evenly convex beneath (Sinaloa, Nayarit, Guerrero). . . . . evansi Gillaspy

23. Tarsal claws not conspicuously paler at base in lateral view, hindfemur weakly and evenly convex beneath as seen in profile (British Columbia, w. U.S., Baja California). .tuberculata (W. Fox)

Tarsal claws, or some of them, conspicuously paler at base in lateral view, hindfemur various. 24

24. Hindfemur straight or slightly concave on basal half beneath as seen in profile, clypeus and tergal bands white (w. U.S. to e. California). . corniculata Mickel

Hindfemur rather evenly convex beneath, clypeal and tergal color various. . . 25

25. Smaller, length 10-13 mm, markings various, labrum ca. 0.66x as broad as long (w. U.S., n. Mexico) . . . . .pulchella (Cresson)

Larger, length 15-19 mm, markings extensively yellow including large scutal U (fig. 16p), labrum ca. 0.85x as broad as long (sw. U.S., n. Chihuahua) . . . . callista J. Parker

Stictiella boharti Gillaspy, (new species) (Fig. 15; map 14)

Male holotype. Length 16 mm, forewing 10 mm. Black, marked with whitish on face (fig. 15d), and yellow as follows: postorbital band ending before inner eye angle; pronotum partly; small posterolateral scutal spots; tegula anteriorly; triangular scutal spots; metanotum; large anterior mesopleural bands meeting ventrally and extending posteriorly to mesocoxa, but not including signum; metapleuron mostly; large V lining propodeal enclosure; sides of propodeum and angles; legs except femora above and tarsomere V apically; terga as shown (fig. 15e); sterna posteriorly; spiracular lobes and S-VII, S-VIII, and gonostylus apically; scape, pedicel, and F-I at base below. Pubescence pale, except brownish on vertex and scutum; appressed silvery on clypeus, frontal orbits, and mesopleuron around signum; some erect hair on clypeus laterally and hindfemur ventrally, hindtibia with a tuft of hair distad (fig. 15o). Punctation of clypeus obscured by pubescence; that of labrum fine, except a few coarse punctures basally. Facial proportions (fig. 15d): galea 1.8x length of labrum and 1.1x eye height, upper clypeal margin separated from antennal sockets by a third of antennocular distance, margin slightly peaked medially, clypeus moderately convex, F-X-XI distinctly flattened and F-X with a strong lateral projection (fig. 15a), midocellar plane 1.1x broader than long, foretarsomere I 7.0x as long as broad, 6 rake setae about as long as article width, II 1.6x basal width, midfemur longitudinally serrate below, midbasitarsus curved and with a basal clump of 4 stout setae, hindfemur with some long hairs below, T-VII spiracular lobes well separated ventrally and with many long bristles along inner edge (fig. 15c), T-VII weakly notched apically, S-II troughlike apically and broadly angulate at margin, S-VIII elongate and medioventrally simple; gonostylus moderately broad, without long hair beneath (fig. 15k, L), cuspis shorter than digitus.

Female. Length 18-19 mm. Yellow markings more extensive than in male and with a pair of discal yellow lines on scutum, midtibia with a dark spot at femorotibial joint; LID (near midocellus) 1.5x eye breadth; scape short, distally broadened; F-I length 1.3x F-II; forebasitarsus 2.9x longer than broad and with 4 well developed preapical rake setae; foretarsomere V slender and 0.9x as wide as IV; head breadth equal to thorax at pronotal lobes; hindfemur angled out medially below (fig. 15i).

California records. Holotype male, Anza, Riverside Co., VII-3-56 (A. Menke, UCD). Paratypes, 6 males, 4 females. INYO CO.: Little Lake, VI-2-07 (C. L. Fox); KERN CO.: Jawbone Canyon (Toms Mill), VI-23-83 (D. Faulkner and Brown), VENTURA CO.: 11 mi. e. Camp Ozena, VII-2-65, on Eriodyction trichocalyx (J. Powell); SAN BERNARDINO CO.: Cajon, VII-8-47 (A. Melander); RIVERSIDE CO.: Morongo Valley, V-25-41 (E. C. Van Dyke); Pinyon Flat, VI-21-41 (E. C. Van Dyke); SAN DIEGO CO.: Borrego Valley, IV-29-55 (M. Wasbauer); San Felipe Creek, VI-23-35 (C. M. Dammers).

Discussion. Stictiella boharti is a member of the pulchella group, in which there is no functional arolium between the unusually long

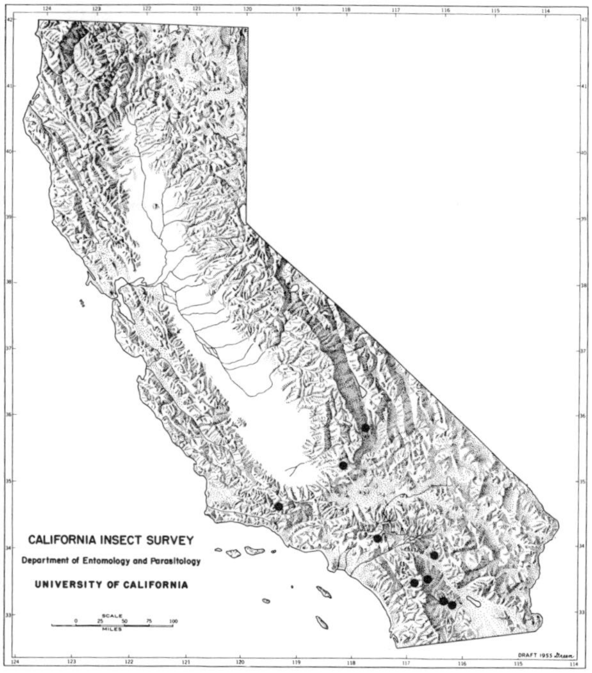

Map 14. California distribution of *Stictiella boharti* Gillaspy.

claws. Others in the group are *callista*, *corniculata*, *evansi*, *pulchella*, and *tuberculata*. The male of *boharti* is unique in its prominent lateral projection on F-X (fig. 15a). The female differs from others in the group by a combination of characters: hindfemur convex below and somewhat angled out medially (fig. 15i), foretarsomeres II-III with contrasting black edges on the posterior corners, head slightly narrower than thorax measured at pronotal lobes, large size, and clypeus as well as terga yellow rather than whitish.

Seasonal distribution is April through July. Flowers visited are *Bebbia* and *Eriodictyon*.

Stictiella callista J. Parker
(Fig. 16; map 15)

Stictiella callista J. Parker 1917:34. Holotype male, "New Mexico" (USNM). According to Cockerell (1918), type locality was Mesilla Park, New Mexico, at flowers of *Chilopsis lineata*.

Geographic range. Desert areas of sw. California, s. Arizona and New Mexico, w. Texas, and n. Chihuahua.

California records. 10 males, 7 females. IMPERIAL CO.: 2 mi. s. Palo Verde, VII-16-62, on *Tamarix* (P. Hurd); Meloland (Experimental Farm), VI-VII, with lycaenid prey (J. Bridwell).

Discussion. *Stictiella callista* is a large (17-19 mm long in the female), mostly yellow wasp belonging to the *pulchella* group, in which the arolium is undeveloped. The male is distinctive by a combination of characters: flagellum simple; S-II with a single dull micropubescent projection (fig. 16r), or sometimes a median hump; S-VI swollen distally and angled back posteriorly (fig. 16q); midbasitarsus distinctly curved (fig. 16f); and foretarsus relatively simple. The female is not so easily characterized: the scutum has a large yellow U (fig. 16p); the clypeus is yellow; the abdomen is nearly all yellow; and the size is greater than in *pulchella*, with yellow forms of which it might otherwise be confused. Bridwell's record of prey was identified by W. D. Field in 1953 as the hairstreak, *Strymon melinus pudicus* Henry Edwards. *S. callista* has been collected on *Chilopsis*, *Tamarix*, *Aster*, *Buddlea*, *Lepidium*, *Lippia*, *Condalia*, and *Prosopis*. We have seen about 50 males and 40 females taken from April to July, but especially in June or July.

Map 15. California distribution of *Stictiella callista* J. Parker. Inset: overall distribution.

Stictiella corniculata Mickel
(Fig. 17; map 16)

Stictiella corniculata Mickel 1918:332. Holotype, Worland, Wyoming (University of Nebraska).

Geographic range. E. California, (map 16), Nevada, se. Oregon, Idaho, Utah and Wyoming.

California records. 13 males, 12 females. INYO CO.: Big Pine, VIII-24-60 (M. Irwin); Deep Springs, IX-10-56, on Chrysothamnus (E. Linsley); 2 mi. e. Lone Pine, on Chrysothamnus X-11-56, X-12-71 (R. Bohart, C. Goodpasture, E. Grissell).

Discussion. Parker (1929) did not treat Stictiella corniculata in his revision, and Gillaspy (1963b) called attention to this discrepancy, along with giving a discussion of distribution and relationships.

S. corniculata belongs to the pulchella group by virtue of the undeveloped arolia, simple male foretarsus, and medioventrally edentate male S-VIII. It is unique in that group by having male S-VI not sharply pointed posteriorly.

The ventrally concave hindfemur of the male simplifies identification in that sex (fig. 17h), but the female is less distinctive. Other features of the male are: F-X flattened and curved (fig. 17a), clypeus and abdominal terga whitish, scutum dark discally or sometimes with small pale spots, distitarsi black-tipped, midtibia incurved posteriorly and flattened distally (fig. 17f), S-II not dentate but somewhat bulging and with much erect pubescence, S-VI posterior margin angled backward but bluntly pointed posteriorly, T-VII ventral spiracular lobes moderately broad (fig. 17c), S-VIII with lateral projections quite short (fig. 17m, n).

Specific characters of the female are: (1) clypeus mostly whitish and twice as broad as long; (2) scutum all dark discally or with pale dots only; (3) femorotibial joints yellow; (4) hindtibia not black-marked inwardly; (5) hindfemur straight, or slightly incurved beneath on basal half; (6) terga with complete and mostly white pale bands; (7) S-II mostly yellow and with black basal mark emarginate medially. Females of pulchella nubilosa and tuberculata are somewhat similar. Character 5 differentiates from p. nubilosa; also, corniculata is a somewhat larger species, the body length about 13-18 mm (males) and 13-17 mm (females). Characters 1, 5, and 6 partly or entirely distinguish from tuberculata; also, the tarsal claws of tuberculata are dark basally.

Map 16. California distribution of Stictiella corniculata Mickel. Inset: overall distribution.

Nesting habits are unknown, but a female from Winnemucca, Nevada, is pinned with a noctuid moth. Chrysothamnus appears to be an exceptionally attractive flower to this species, and collecting records from July to October parallel its flowering period. However, it has been taken also on Tetradymia and Melilotus. We have seen about 35 males and 20 females.

Stictiella emarginata (Cresson)
(Fig. 18; map 17)

Monedula emarginata Cresson 1865:468. Lectotype female (Cresson Jr. 1916), "Colorado" (ANSP).

Monedula mamillata Handlirsch 1890:146. Lectotype male (Bohart 1970), "Georgie" (Geneva Museum).

Stictiella fairchildi Bohart 1982:593. Holotype male, Gainesville, Florida (UCD). New synonymy.

Geographic range. Mostly in the Canadian Life Zone, occasionally in the Transition, Upper Sonoran, or Austroriparian life zones;

widespread in the United States; also in Alberta, Canada, and Baja California (map 17).

California records. SISKIYOU CO.: 3 mi. n. Medicine Lake; MONO CO.: White Mts. (10,000 ft.), Leavitt Meadows; INYO CO.: Bishop, Big Pine Creek (8,000 ft.); TULARE CO.: Giant Forest, Whitney Portal, Kings Canyon; SAN LUIS OBISPO CO.: 5 mi. ne. Santa Margarita; SAN BERNARDINO CO.: Big Bear Lake, Barton Flats; RIVERSIDE CO.: Santa Rosa Mt.

Discussion. *Stictiella emarginata* is a widespread and variable species. It belongs to the *emarginata* species group, in which the arolium is small, especially in the female. Species features in both sexes are: hindfemur with abundant, erect hair beneath (at least in western material, and particularly in males) (fig. 18h); flagellum all dark, or with a small pale spot at base of F-I; labrum at least as long as F-I-II together; frons considerably broader then eye (fig. 18d); and S-II with a large black area medially (fig. 18o). Males and most females (see exceptions below) have the scutum all dark discally. Males have the projections of S-II and the medioventral spine of S-VIII unusually stout (fig. 18o, m); the spiracular lobes of T-VII are moderately broad (fig. 18c).

In the Santa Rosa and Ruby Mts. of Nevada, and in the Rocky Mountain region of New Mexico (Portales), Utah, Colorado, Wyoming, and Montana, there is a strong tendency toward whitish dorsal markings. The lectotype from "Colorado" fits into this category. On the other hand, more western and eastern individuals have yellow markings, including all we have seen from California. Furthermore, some yellow-marked specimens occur with white ones in such localities as Pinedale, Sublette Co., Wyoming, and in the mountains of Weber Co., Utah.

Another variation in markings occurs in females from Arizona. Specimens from Apache Junction, Pinal Co., and Kayenta, Navajo Co., have a nearly complete yellow scutal U. Also, s. Utah females (Kents Lake, Beaver Co.) may have long scutal dashes, as do some from e. New Mexico (Portales, Roosevelt Co.). Material from southeastern U.S. and Baja California offers still another variation in the partial reduction of erect hair beneath the hindfemur. *S. emarginata* is a relatively large species with body length usually 13-16 mm in males, and 13-19 mm in females. However, we have seen some males as short as 11 mm.

Nesting has not been observed in this species. Bradley (1908) reported sleeping aggregations. Gillaspy et al. (1962:562) recorded a specimen pinned with a moth of the genus *Euxoa* (Noctuidae) from the White Mountains of Mono Co., California. The U.C. Davis Collection has 4 females pinned with noctuids from mountainous areas of Mono, Inyo, Alpine, and Tulare counties of California.

Most specimens are collected during June through August, but we have seen females collected in Arizona in May (Kayenta) and April (Apache Junction). We have seen about 125 males and 155 females.

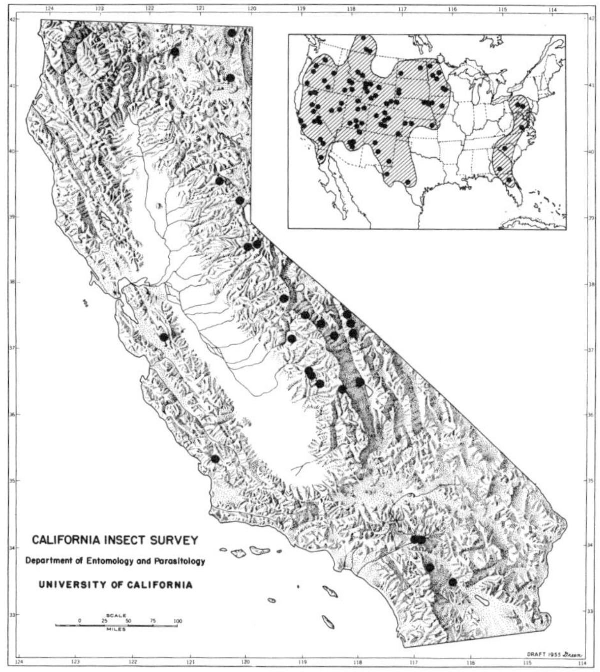

Map 17. California distribution of *Stictiella emarginata* (Cresson). Inset: overall distribution.

Stictiella fergusoni Bohart, (new species)

(Frontispiece; fig. 19; map 18)

Holotype male. Length 18 mm, forewing length 11 mm. Black, marked with whitish and yellow. Whitish are: clypeus and lower frons laterally; tergal markings except laterally (fig. 19e); gonostylus on apical half, except outer brown strip (fig. 19k). Yellow are: labrum; midocellar mark (fig. 19b); pronotum mostly; narrow scutal strip alongside tegula; scutellar and metanotal bands; propodeum, except basal and submedial posterior bands;

mesopleuron mostly (black spot covers signum); legs mostly, except black dorsoposterior spots on femora and black tarsomere V, which is translucent medially on foreleg; sternum mostly, except some basal spots and long oval spot on S-II (fig. 19n). Pubescence pale, inconspicuous, hindfemur without obvious hair beneath. Punctation fine and inconspicuous, a few scattered punctures on labrum. Face (fig. 19d) with galea shorter than eye height and 1.2x as long as labrum, F-XI 3x as long as broad (fig. 19a), clypeus evenly convex, midocellar flat area as long as broad, foretarsus (fig. 19g) with articles tapering in width to greatly flattened and expanded tarsomere V, forebasitarsus 5x as long as broad and with 6 short rake setae, claws slender and equal, arolium moderately developed, midtarsus (fig. 19f) with basal tarsomere curved, midfemur distinctly serrate on distal three-fourths (fig. 19i), hindfemur weakly convex beneath, T-VII not indented at apex, spiracular lobes broad and narrowly separated medioventrally (fig. 19c), S-II weakly bituberculate (fig. 19n), S-VI somewhat longitudinally ridged but posterior margin rounded, S-VIII without a medioventral tooth (fig. 19m), gonostylus narrow but rounded distally (fig. 19k, L).

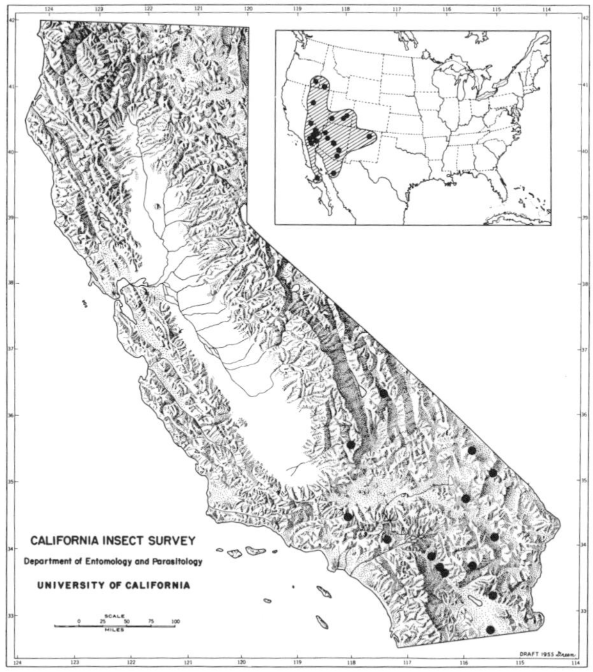

Map 18. California distribution of Stictiella fergusoni Bohart. Inset: overall distribution.

Female. Size and markings about as in male, except scutum with a well defined yellow scutal U, tarsi yellow, and terga with more extensive and all-yellow banding. Forebasitarsus nearly 3x as long as broad and bearing 5 well developed preapical rake setae (fig. 19j).

Geographic range. Lower and Upper Sonoran life zones in s. California, se. Oregon, sw. Idaho, and Utah, Arizona, New Mexico, and nw. Mexico (map 18).

Records. Holotype male, Granite Mts. (20 mi. s. Kelso), San Bernardino Co., California, X-10-77, on Chrysothamnus (R. M. Bohart, UCD). Paratypes, 43 males, 23 females. California: INYO CO.: Darwin Falls, X-10-76 (T. Griswold); KERN CO.: Sand Canyon w. of Inyokern, X-11-77 (R. Bohart); SAN BERNARDINO CO.: 25 mi. s. Ivanpah, X-12-58 on Chrysothamnus (R. Thorp); Mexican Well near Mountain Pass; New York Mts., IX-24-71 (C. Goodpasture); Granite Mts., X-10-77 (R. Bohart); 6 mi. ne. Chubbock, X-15-71 (E. Linsley, P. Hurd); Victorville, IX-21-38 (P. Timberlake); LOS ANGELES CO.: Palmdale, X-6-35 (E. Linsley); RIVERSIDE CO.: Whitewater, X-5-61 (collector?); Joshua Tree Monument, X-13-61 (E. Sleeper); Andreas Canyon near Palm Springs, X-7-77 (N. Smith); Deep Canyon near Palm Desert, X-8-77 (R. Bohart); IMPERIAL CO.: 11 mi. e. Beal Well, X-24-51, on Hyptis (P. Hurd); Salton Sea, X-20-63 (L. McDermid). Oregon: HARNEY CO.: Alvord Desert, IX-6-79 (T. Siebert), VIII-IX, 1979 (Neil Cobb, T. Siebert, M. Smith). Idaho: OWYHEE CO.: Hot Creek Falls, IX-9-65 (W. Barr). Utah: EMERY CO.: Goblin Valley, IX-16-79 (F. Parker, D. Veirs); near Little Gilson Butte, IX-17-80 (T. Griswold), VIII-26-81 (F. Parker, T. Griswold); WAYNE CO.: Capitol Reef, IX-16-81 (C. Hatley, G. Briggs); Hanksville, VIII-31-82 (R. Bohart); WASHINGTON CO.: Beryl, IX-21-? (G. Knowlton). Nevada: CHURCHILL CO.: Sand Mt., IX-5-79 (R. Bechtel, R. Bradley). Arizona: PINAL CO.: Apache Junction, X-25-58 (R. Dawson); Phoenix, XI-3-54 (G. Butler); MOHAVE CO.: Kingman, X-16-73 (P. Torchio, F. Parker), Wickiup, IX-11-57 (J. Gillaspy); PIMA CO.: Three Points, IX-17-68 (E. Erickson); Sabino Canyon; Madera Canyon. New Mexico: TAOS CO.: Rinconada. Mexico: BAJA CALIFORNIA: 20 mi. n. Mesquital, IX-27-41 (E. Ross, G. Bohart); SONORA: Hermosillo.

Discussion. Stictiella fergusoni is closely related to speciosa (Cresson), which in general has a more easterly distribution.

S. speciosa occurs from Nebraska to Texas and n. Mexico westward through Colorado to Arizona, where the ranges overlap. S. fergusoni differs as follows: male forefemur nearly glabrous beneath (instead of distinctly pubescent), foretarsomere V unusually broad and medially translucent (fig. 19g); female fergusoni with hindtarsus yellow rather than darkened toward apex; terga yellow-banded rather than somewhat infused with white.

Flowers visited are Chrysothamnus, Dicraurus, Aster, Helianthus, Bidens, Argemone, Solidago, Viqueriera, Hyptis and Baccharis.

This species is named for my friend George R. Ferguson, who has contributed significantly to our knowledge of philanthine wasps.

Stictiella flavescens Gillaspy, (new species) (Fig. 20; map 19)

Male holotype. Length 15 mm, forewing 9 mm. Black, extensively marked with yellow as follows: face (fig. 20d), posterior orbital bands continuous behind eyes, pronotum entirely, scutum laterally and a broad discal U (as in fig. 20h), anterior half of tegula, broad transverse bands on scutellum and metanotum, propodeum mostly; pleuron and body ventrally nearly all yellow, except narrow basomedial black on S-V-VI; terga broadly (fig. 20e); scape yellow, except small apical dot above; pedicel and flagellomeres slightly brownish-yellow below, brown above. Pubescence pale, inconspicuous, no silvery appressed type, clypeus without and hindfemur practically without erect or semierect hair. Punctation exceptionally fine and dense on clypeus, fine to medium fine elsewhere.

Facial proportions (fig. 20d): LID 0.32x head breadth, face widening only slightly above, galea 1.6x as long as labrum and 0.9x eye height; clypeus 0.4x socket diameter below socket, antennocular distance 1 socket diameter, upper clypeal margin slightly concave; foretarsus slender, forebasitarsus 6.0x as long as broad, with 6 rake setae as long as tarsomere width, V 3.5x as long as broad, arolium about one-third of claw length; midfemur serrately carinate beneath (fig. 20i), midbasitarsus curved and with a clump of bristles near base within; T-VIII with basal glabrous area angularly extended posteriorly at midline (fig. 20f), with broad spiracular lobes nearly touching ventrally (fig. 20c); S-II without processes or evident tumescences; S-VI bending outward (downward), troughlike toward apex; S-VII almost as long as T-VII; S-VIII with 3 slender prongs, but no basomedian spine or carina (fig. 20m); gonostylus (fig. 20k, L).

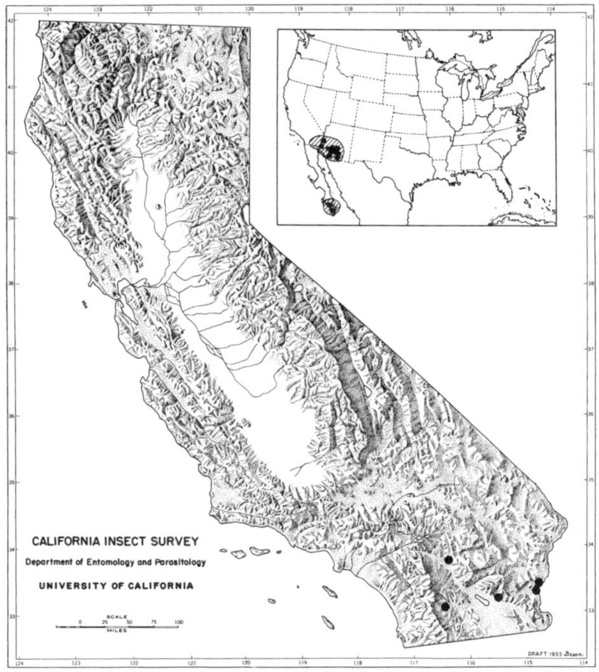

Map 19. California distribution of Stictiella flavescens Gillaspy. Inset: overall distribution.

Female. Length 14mm. Similar to male, a little silvery appressed pubescence along inner orbits, LID about 0.38x head breadth, foreleg arolium quite small.

Geographic range. Mostly Lower Sonoran Life Zone localities in se. California, s. Arizona, ne. Sonora, and Baja California Sur.

Records. Holotype male, Scissors Crossing, San Diego Co., California, VIII-4-61 on Petalonyx thurburi (J. E. Gillaspy, UCD). Paratypes, 31 males, 22 females. California: RIVERSIDE CO.: Thousand Palms; Ripley, VIII-19-46 (P. Hurd). SAN DIEGO CO.: Scissors Crossing, VIII-4-61, on Petalonyx and Heliotropium (J. Gillaspy); IMPERIAL CO.: Palo Verde, VIII-15-46 (P. Hurd), X-10-71 (R. Bohart); 11 mi. e. Beal Well, X-24-51 (P. Hurd). Arizona: MARICOPA CO.: Sauceda Mts. (20 mi. s. Gila Bend), VII-27-31 (E. Tinkham); PINAL CO.: Picacho Pass, VIII-7-40 (W. Swisher); YUMA CO.: Dateland, VIII-11-54 (G.

Bohart and G. Butler); near Yuma, IV-3-57 (T. Haig); PIMA CO.: near Ajo, VII-2 to VII-27-61, on Condalia and Petalonyx (J. Gillaspy); Avra Valley, VII-6-55 (G. Butler, F. Werner); Childs, VII-26-61, on Tamarix (J. Gillaspy); 9 mi. w. Sells; Tucson, VII-14-55 (G. Butler). Mexico: BAJA CALIFORNIA SUR: 10 mi. nw. El Triunfo, on Asclepias (E. Linsley); near La Paz, IX-3-59 (K. Radford, F. Werner); VIII-2 to VIII-4-66, on Wislizenia (P. Hurd, J. Chemsak); SONORA: 20 mi. sw. Sonoyta, VI-13-52 (M. Cazier et al.).

Discussion. Stictiella flavescens is quite similar in size and general appearance to pulchella, but differs in both sexes by the narrower face and definite, though small, arolium. The male has additional differences in the smooth rather than tuberculate S-II, and also in the longer S-VII which more completely closes the apex of segment VII. There is a superficial resemblance to villegasi, based on the extensive yellow coloration. However, flavescens has the labrum longer than F-I-II together, and in the male S-VIII is simple basomedially.

Stictiella pulchella (Cresson)

This is a polytypic species and, together with emarginata, one of the two Stictiella which range across the continent. As the nominal species of the pulchella group, the arolia are undeveloped, and male characters are: foretarsal V slender (fig. 21g), midbasitarsus curved (fig. 21f), S-VI angled back and pointed, and S-VIII edentate medioventrally. Additionally, in both sexes of pulchella the hindfemur is rather evenly convex beneath in profile, and the tarsal claws are conspicuously paler basally. Male S-II is simple or weakly bidentate, and the flagellum is simple (fig. 21a).

The male can be separated from tuberculata by the simple foretarsus, from corniculata by the convex hindfemur, from evansi by the curved midbasitarsus, from boharti by the simple antenna, and from callista by the smaller size (body length 8-14 mm), and the simple or bidentate S-II.

The female, as in most Stictiellina, is less distinctive. It can be separated from evansi and boharti by the all-yellow femorotibial joints, from tuberculata by the pale base of the tarsal claws, from corniculata by the more evenly convex hindfemoral outline beneath, and from callista by the smaller size (body length 9-13 mm) and longer labrum (about two-thirds as broad as long).

We are recognizing 4 subspecies: pulchella, nubilosa, plana, and serrata. These names have been used previously to designate distinct species, but we have been unable to find significant differences other than markings. On the other hand, these forms are fairly distinct geographically, even though some intermediates can be found. Only S. pulchella and S. nubilosa are Californian. S. p. plana occurs in the Great Plains states, and S. serrata from North Carolina to Florida.

KEY TO SUBSPECIES OF STICTIELLA PULCHELLA

1. Body extensively yellow, scutum in male usually with long yellow dashes discally and sometimes nearly complete scutal U, females usually with well developed scutal U (fig. 21a) (sw. U.S., n. Mexico). . . . . pulchella pulchella (Cresson)

   Body less extensively yellow, sometimes white-marked, scutum in male dark discally, scutum in female with short pale discal dashes at most. 2

2. Abdomen ventrally with complete pale yellow bands, those of female infused with whitish, female clypeus all pale (sw. California, w. Nevada, e. Washington). . pulchella nubilosa Gillaspy

   Abdomen ventrally with few, if any, complete pale bands, tergal bands yellow or whitish, female clypeus often with a bilobed black spot above 3

3. Black marks on legs not tinged with reddish; bands on T-III and T-IV usually entire (U.S.: Great Plains States, Mexico: Veracruz) . . . . pulchella plana (W. Fox)

   Black marks on legs tinged with reddish; bands on T-III and T-IV usually interrupted medially (se. U.S.). . . . . .pulchella serrata (Handlirsch)

Stictiella pulchella pulchella (Cresson), (new status) (Fig. 21; map 20)

Monedula pulchella Cresson 1865:471. Lectotype female (Cresson Jr. 1916), "Colorado" (ANSP).

Stictiella melanosterna J. Parker 1917:30. Holotype male, "New Mexico" (USNM). According to Cockerell (1918), type locality was Las Cruces, New Mexico.

Geographic range. S. California in the east and south, and east through Arizona, s. Colorado, and New Mexico to w. Texas. It extends into n. Mexico: Sonora, Chihuahua, Coahuila, and Baja California. Most localities are in the Upper and Lower Sonoran life zones.

California records. INYO CO.: 3 mi. n. Lone Pine, w. side Owens Lake, Keeler, Olancha; SAN BERNARDINO CO.: 9 mi. s. Baker, 2 mi. s. Kelso, Cronise Valley, Barstow, Victorville, Morongo Valley; RIVERSIDE CO.: Thousands Palms, Palm Springs, Cathedral City, Mecca, Indio, 18 mi. w. Blythe, Ripley, Anza; SAN DIEGO CO.: Borrego Valley, Scissors Crossing, Warner Springs, 14 mi. e. Julian, Jacumba; IMPERIAL CO.: Palo Verde, 17 mi. nw. Glamis, 11 mi. e. Beal Well, Meloland.

Discussion. The typical subspecies is relatively abundant in desert areas of s. California. Both sexes are extensively yellow, particularly the females, which superficially resemble Stictiella callista, S. flavescens, Glenostictia scitula, and Xerostictia longilabris. The color pattern, highlighted by the long yellow, scutal dashes of the male and well developed scutal U in the female (fig. 21j), is remarkably consistent in the Lower Sonoran Life Zone of Riverside, Imperial, and San Diego counties. However, as the range extends into the Upper Sonoran of more upland areas, such as Banning and Warner Springs, some intermediates of both sexes can be found between pulchella s.s. and p. nubilosa. The lectotype female of pulchella from "Colorado" is itself a little intermediate, since the scutal "U" is fragmentary. Material we have seen from n. central Colorado (Roggen, Weld Co.) is obviously subspecies plana. Occasional males from within the range of pulchella s.s. may have the yellow scutal dashes totally suppressed or almost so. We have seen a few of these from White Sands and Las Cruces, New Mexico, and from Blythe, California.

Information on the biology of pulchella s.s. (as melanosterna) was given by Gillaspy in Gillaspy et al. (1962:562). Clustering at night was seen near Bingham, New Mexico, and near Lajitas, Brewster Co., Texas. Nesting was observed in a sandy area at the latter site. The nests were 12-14 cm deep and unicellular. The tumulus is apparently scattered. Cells were stocked with moths of the genera Loxostege (Pyralidae) and Characoma (Noctuidae). The egg was laid on the last prey in the cell, as usual in mass provisioning. Evans (1966:182) presented some evidence that nests may sometimes have 2 cells. Gillaspy et al. (1962:559) also reported additional prey as a butterfly, Leptotes (Lycaenidae). Evans (1964) described the larva.

A large variety of flowers are visited during the flight period of May to November. Most collections have been made from May to August. We have studied about 1,200 specimens of both sexes.

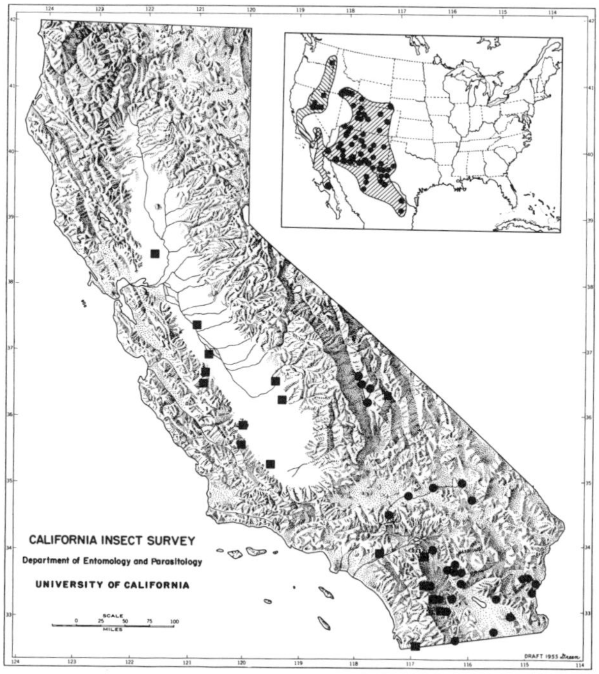

Map 20. California distribution of Stictiella pulchella pulchella (Cresson) (black circles) and S. pulchella nubilosa Gillaspy (black squares). Inset: overall distribution.

Stictiella pulchella nubilosa Gillaspy, (new status) (Map 20)

Stictiella nubilosa Gillaspy 1963b:252. Holotype female, "Los Angeles Co., Cal." (USNM).

Geographic range. Southern two-thirds of California in the w. half and in the Upper Sonoran Life Zone. The range extends into w. Nevada, and north to e. Washington (map 20).

California records. YOLO CO.: Davis; STANISLAUS CO.: Turlock; MERCED CO.: Dos Palos; SAN BENITO CO.: near Panoche Pass; FRESNO CO.: Olema, Panoche Road, Los Gatos Canyon (Diablo Mts.); TULARE CO.: Goshen Junction; KINGS CO.: Kettleman City; KERN CO.: Buttonwillow; SAN BERNARDINO CO.: Colton; RIVERSIDE CO.: Banning, Anza; SAN DIEGO CO.: Warner Springs, Scissors Crossing, Agua Caliente.

Discussion. As indicated in the key, subspecies *nubilosa* has the scutum all dark or almost so in the male, and at most weakly spotted in females. The male has pale yellow tergal bands which are complete on T-II to T-VII but usually broadly broken on T-I. The female has the tergal bands infused with whitish and usually complete on T-I-VI. Both sexes have the abdomen extensively pale. The hindtibia frequently has a black spot or longitudinal dash on the inner surface. This is absent in *pulchella pulchella*, but sometimes present in the other subspecies.

Adults have been collected at flowers of *Petalonyx*, *Croton*, *Eriogonum*, and *Solidago* in the months of June to September. We have studied about 500 specimens of both sexes.

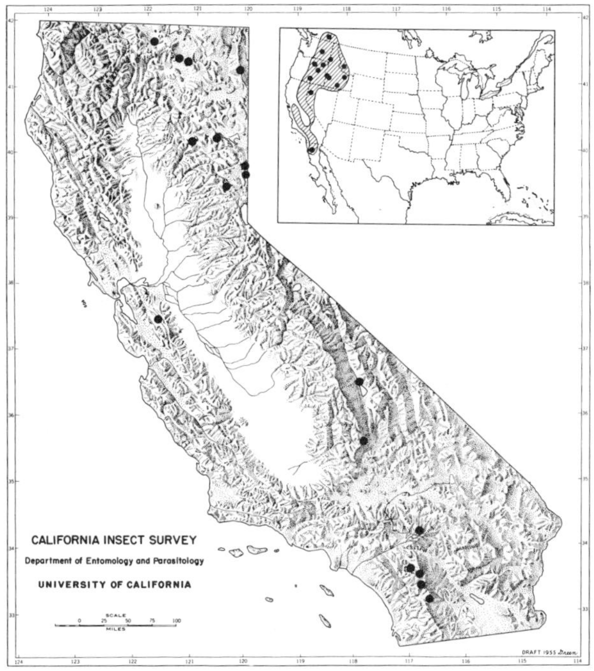

Map 21. California distribution of *Stictiella tuberculata* W. Fox. Inset: overall distribution.

Stictiella tuberculata (W. Fox)
(Fig. 22; map 21)

Monedula tuberculata W. Fox 1895:366. Holotype male, "Nevada" (ANSP).

Geographic range. Upper Sonoran Life Zone hills and low mountains in California, Nevada, Oregon, Washington, Idaho, Montana, Baja California, and British Columbia.

California records. SISKIYOU CO.: Macdoel; MODOC CO.: Lake City, 3 mi. s. Perez, 10 mi. s. Canby; PLUMAS CO.: Chester; LASSEN CO.: Susanville, Constantia, Hallelujah Junction; SIERRA CO.: Sattley; ALAMEDA CO.: Tesla; INYO CO.: Lone Pine; KERN CO.: Walker Pass; RIVERSIDE CO.: Gavilan, Hemet Reservoir, Anza; SAN DIEGO CO.: 2 mi. n. Warner Springs.

Discussion. *Stictiella tuberculata* is unique in the *pulchella* group through having foretarsal V in the male expanded and blackened (fig. 22g). Another outstanding feature of the male is the rounded and densely pubescent medioventral projection of S-II (fig. 22i). Other specific characters of the male are: (1) clypeus whitish and somewhat blackened above (fig. 22d); (2) scutum all or nearly all black; (3) hindfemur nearly straight beneath in middle three-fifths; (4) T-VII somewhat contorted, and its spiracular lobes moderate (fig. 22c); (5) 3 prongs of S-VIII well developed. Pale tergal markings may be somewhat whitish, and they occasionally form complete bands even on T-I. This circumstance, along with the black-tipped tarsi, pubescent S-II, partly white clypeus, and moderate T-VII spiracular lobes creates a slight similarity to *corniculata*. However, foretarsal V is expanded in *tuberculata*, and the pubescent knob on its S-II is distinctive. Other characters in the list above which do not agree wholly or in part are 1, 3, 4 and 5.

Female characters are: (1) clypeus yellow, usually margined with black above (fig. 22h), about twice as broad as long; (2) scutum all dark discally, or with small yellow spots; (3) femorotibial joints yellow; (4) hindfemur rather evenly convex beneath, (5) hindtibia all yellow inwardly; (6) tergal yellow bands sometimes incomplete on T-I; (7) S-II black basal mark emarginate medially. *S. corniculata* is a somewhat similar species, but it does not agree wholly or in part with characters 1 (color), 4 and 6. *S. pulchella* has some similarity also, but it is a much more extensively yellow species.

Nothing is known about nesting habits. Adults fly from May to October, and visit flowers of *Eriogonum*, *Chrysothamnus*, *Cryptantha*, *Croton*, *Achillea*, and *Melilotus*. We have seen about 30 males and 10 females.

Stictiella villegasi Bohart
(Fig. 23; map 22)

Stictiella villegasi Bohart 1982:596. Holotype male, Glamis, Imperial Co., California (UCD).

Geographic range: Se. California in desert areas of the Lower Sonoran Zone (map 22).

California records. 7 males, 8 females. IMPERIAL CO.: Glamis, X-11-72 (B. Villegas, C. Goodpasture); 3 mi. nw. Glamis, X-16-72 (M. Wasbauer, E. Hardy); Holtville, XI-25-55 (C. Dammers); 20 mi. w. Yuma, X-17-75 (P. Torchio, F. Parker); Yuma Sand Dunes, X-17-76 (S. McElfresh, D. Faulkner).

Discussion. This rare species is almost entirely yellow and bears a superficial resemblance to *Xerostictia longilabris*. The diagnostic characters are: arolium distinct (fig. 23g), male midbasitarsus curved (fig. 23f), hindfemur with only a few small hairs beneath, wing membrane clear, labrum unusually short (fig. 23d), and female F-I as long as scape.

Other features of note in the male are the relatively simple antenna, with F-I about as long as F-II (fig. 23a); a nearly complete yellow scutal U (as in fig. 23h); midfemur rather weakly serrate (fig. 23i); S-II weakly bidenticulate (fig. 23j); spiracular lobes narrow (fig. 23c); and S-VIII with a medioventral projection (fig. 23m).

Genus *Microstictia* Gillaspy

Monedula, *Stictiella*, *Stictia*, *Glenostictia* of authors, in part.
Microstictia Gillaspy 1963a:196. Generotype: *Monedula femorata* W. Fox, original designation.

The genus *Microstictia* contains 10 species, which are limited to the Nearctic Region and for the most part to areas west of the 100th meridian. There are 4 species found in California, all in east-central and southeastern parts of the state.

Biology. Little has been published previously on nesting biology. Gillaspy (1983)

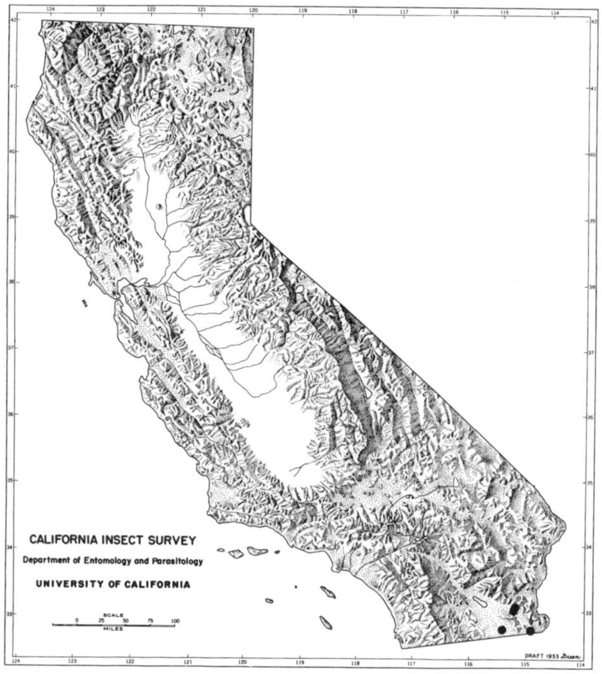

Map 22. California distribution of *Stictiella villegasi* Bohart.

observed activities of *M. texensis* at Kingsville, Texas, during the summers of 1967-69. Nests were constructed on the banks of an intermittent stream in huisache-granjeno-mesquite-prickly pear brush. Whining, hovering, and occasionally even backing flight were characteristic. Provisioning wasps approaching their nest with prey threaded their way noisily and rather slowly through vegetation, but departed abruptly with a burst of sound toward open areas. Wasps were timed during 77 trips away from the nest, which averaged 12 minutes (range 3-47); only 5 returns were without prey. Time in the nest for prey storage (n = 28) averaged 24 seconds (range 15-45). Tumuli were not dispersed, and while temporary closures were made frequently, they were usually omitted during the most active periods of provisioning. Burrows were 8-19 cm in depth, 6-7 mm in diameter, and had 1-7 cells. Either intercepted or removed from cells were 31 adult moth prey which included Gelechiidae, Geometridae, Olethreutidae, Pterophoridae, and Pyralidae (Crambinae, Phycitinae, Pyraustinae). Prey moths had the forewing base and scutum notably denuded of scales, apparently through carrying action of the wasp, but were otherwise intact, and one of those taken from a cell exhibited vigorous heart action when viewed by stereoscopic microscope. Paralysis otherwise appeared complete. One closed cell had 11 prey. The 2.3 mm long wasp egg in another cell was glued to the hindwing of the first moth stored, near the base of the left hindleg. Flowers visited were Phyla incisa (often), Heliotropium curassavicum, Ratibida columnaris, Aster ericoides, Borrichia frutescens, and Croton sp.

Adults of other species have been collected at flowers such as Eriogonum, Bebbia, and Hyptis. Both sexes hover and produce the whine-like sound noted above for texensis, this is even higher pitched than in Glenostictia or Steniolia. Were it not for this sound, it would be difficult for the observer to spot them in the field.

Taxonomy. Bohart and Menke (1976:550) merged Microstictia and Stictiella. With the synonymy of femorata and exigua, there were only 3 species of Microstictia named at that time, and these had the male midfemur modified somewhat as in Stictiella. Since 10 species are now available, and all of these are easily recognized by the structure of the midocellar area, we are considering Microstictia to be a distinct genus. An additional point is that in 4 of the recently described species in which males are known, the midfemur is unmodified in this sex.

Characters of the genus are as follows: (1) midocellar plane forming a semicircle or ellipse which is about 2x as broad as long (fig. 13b); (2) size small, with body length rarely exceeding 12 mm; (3) labrum nearly flat in lateral profile; (4) palpal formula 6-4; (5) clypeus gently convex in lateral profile, and straight to concave along upper medial margin; (6) arolium developed, but small in female; (7) hindfemur without erect hair beneath; (8) male midfemur smooth or irregular beneath, but not serrate; (9) male S-II with a double projection which is sometimes on a stalk; (10) male S-VIII with a medioventral tooth or raised carina. On the basis of the male midfemur, Microstictia can be divided into 2 groups:

1. The femorata group. Male midfemur compressed beneath into a biangulate carina or lamina. Included species: femorata, flavida, hurdi, minutula, nova (presumptive, female only known), texensis.
2. The ocellata group. Male midfemur smooth and rounded beneath. Included species: deserticola, lobata, ocellata, rufescens.

KEY TO SPECIES OF MICROSTICTIA

1. Flagellum with 11 articles, abdomen with 7 terga (males). 2

   Flagellum with 10 articles, abdomen with 6 terga (females). 10

2. Midfemur unmodified, midtarsomere V all pale. 3

   Midfemur modified by posteroventral emarginations and/or sharp angles, midtarsomere V often with dark spot distally beneath. 6

3. S-II with 2 separated toothlike projections, T-VII spiracular lobes narrow (figs. 24n, 26h). 4

   S-II with single median projection which is double at apex (fig. 25n), T-VII spiracular lobes various. 5

4. Fore- and midfemora all yellow or nearly so, S-VII with broad median sclerotized area (fig. 24c) (s. California, s. Arizona). . . . . . . deserticola Bohart

   Fore- and midfemora with large black spot posteriorly, S-VII with moderately narrow median sclerotized area (fig. 26c) (e. California, Nevada, Arizona). . . . . ocellata Bohart

5. Whitish tergal bands usually tinged with reddish anteriorly and posteriorly, but not yellow laterally; sternal bands whitish; dark spot above on hindtarsomere V; lateral prongs of S-VIII black; T-VII spiracular lobes narrow (s. California to Texas, Nuevo Leon, Baja California Sur). . . . . .rufescens Gillaspy

   Whitish tergal bands not tinged with reddish, but yellowed laterally; sternal bands yellow; hindtarsomere V all pale; lateral prongs of S-VIII mostly whitish; T-VII spiracular lobes somewhat broadened (fig. 25c) (e. and s. California, s. Arizona)..lobata Bohart

6. Basoventral swelling of midfemur rounded or obtusely angulate. 7

Basoventral swelling of midfemur sharp and acute in profile. 8

7. Basoventral swelling of midfemur angulate, but obtuse in profile; S-VIII with only a basoventral ridge (Texas, Nebraska). . . . minutula (Handlirsch)

Basoventral swelling of midfemur rather weak and simply rounded in profile, S-VIII with a basomedian and retrorse tooth (Arizona, New Mexico, Sonora, Chihuahua, Oaxaca). . flavida Gillaspy

8. Foretarsus hardly maculate beneath, midtarsomere V without dark spot apically beneath (Arizona, New Mexico, Sonora, Sinaloa, Oaxaca). . . . .hurdi Gillaspy

Foretarsus conspicuously maculate beneath, midtarsomere V with dark spot apically. 9

9. Posteroventral concavity of hindfemur extending a little more than two-thirds of femoral length (w. Texas, Kansas, New Mexico, Colorado, Arizona, Chihuahua, Durango; Florida--probably in error). . . . . . . femorata (W. Fox)

Posteroventral concavity of hindfemur extending a little less than two-thirds of femoral length (e. Texas, Coahuila, Tamaulipas). . . . . .texensis Gillaspy

10. Pale bands of abdominal terga III-IV whitish dorsally and laterally. . . .11

Pale bands of abdominal terga III-IV either all yellow or at least yellow laterally. 13

11. Pale tergal bands edged in front and behind with reddish, clypeus white with black spots or band above (s. California to Texas, Nuevo Leon, Baja California Sur). . . . . rufescens Gillaspy

Pale tergal bands not edged with reddish, clypeus without large black spots or band. 12

12. West of 100th meridian: w. Texas, New Mexico, Kansas, Colorado, Arizona, Chihuahua, Durango. .femorata (W. Fox)

East of 100th meridian: s. half of Texas, Tamaulipas, Coahuila. . . . . texensis Gillaspy

13. Clypeus with labrum whitish. 14

Clypeus and labrum yellow. 15

14. Forefemur and midfemur all yellow posteriorly, or with small black spots (s. California, s. Arizona) .deserticola Bohart

Forefemur and midfemur with large black stripes posteriorly (e. and s. California, s. Arizona). . . .lobata Bohart

15. Tergal pale bands white, except yellow laterally; scutum all dark or with small pale dashes discally (e. California, Nevada, Arizona). . . . . ocellata Bohart

Tergal pale bands yellow, sometimes a bit paler toward middle; scutum usually with large yellow dashes or a nearly complete U. 16

16. East of 108th meridian in Montana, Nebraska, and s. into Texas, tergal bands usually paler medially. . . . . minutula (Handlirsch)

West of 108th meridian, tergal bands usually all yellow. 17

17. Galea broad, ending abruptly in an obtuse point, 0.7x as long as eye height; yellow band across thoracic venter connected to pleural yellow (Arizona). . . . . nova Gillaspy

Galea narrow and tapering to a narrowly rounded point, 0.9x as long as eye height, thoracic venter various. . . 18

18. Mesopleural yellow areas separated ventrally by black, facial black marks extending downward to antennal sockets (Arizona, New Mexico, Sonora, Sinaloa, Oaxaca). . . . . hurdi Gillaspy

Mesopleural yellow areas broadly connected ventrally, facial black marks not extending downward to antennal sockets (Arizona, New Mexico, Sonora, Chihuahua, Oaxaca). . flavida Gillaspy

Microstictia deserticola Bohart, (new species) (Fig. 24; map 23)

Male holotype. Length 8 mm, forewing 5.5 mm. Black with yellow as follows: outer orbital line broken medially, prothorax except basally, scutum laterally and wing base, a pair of scutal dots, scutellar band narrowed medially, metanotum, posterior band on propodeal enclosure, propodeal angle, pleuron almost entirely, legs entirely except small dark mark behind metafemur (other femora with tiny black dots in some paratypes), broad tergal bands becoming a little lighter medially (fig. 24e), S-I-VI almost wholly, apical parts of S-VIII and gonostylus. Facial markings are whitish (fig. 24d), flagellomeres all very light reddish-yellow beneath. Pubescence pale, silvery appressed on clypeus, lower frons, and to some extent in many other areas. Punctation fine and inconspicuous. Facial proportions (fig. 24a, b, d, i, j): galea 1.7x as long as labrum and 0.9x eye height; F-XI 2x as long as broad (fig. 24a); forebasitarsus 5x as long as broad, with 5 rake setae as long as tarsomere width (fig. 24g); midfemur and hindfemur unmodified; T-VII spiracular lobes simple and narrow; S-VII with a broad median sclerotized area (fig. 24c); S-VIII with a strong medioventral tooth (fig. 24m); gonostylus (fig. 24k, L).

Female. Length 8-8.5 mm, forewing 5.5 mm. Markings as in male, but yellow more extensive. Scutum with a nearly complete discal U. Abdomen dorsally pale yellow to whitish, deeper yellow laterally. Forebasitarsus 2.6x as long as broad, with 4 well developed preapical rake setae. Arolia small (fig. 24h).

Geographic range. Lower Sonoran localities in s. California and w. Arizona (map 23).

Records. Holotype male, Westmorland, Imperial Co., California, VI-31-30 (P. H. Timberlake, CAS). Paratypes, 3 males, 2 females. *California*: SAN BERNARDINO CO.: Colton Hills, VI-17-80 (T. Griswold); Cronise Valley, VII-9-38, on *Heliotropium* (P. Timberlake); RIVERSIDE CO.: Hopkins Well, IV-29-52, on *Heliotropium* (P. Hurd); SAN DIEGO CO.: Borrego Valley, IV-29-54, on *Bebbia juncea* (M. Wasbauer). Other paratypes, 1 male, 1 female. *Arizona*: PINAL CO.: 10 mi. s. Toltec, VI-21-53 (T. Haig); YUMA CO.: Roll, VI-2-55 (G. Butler).

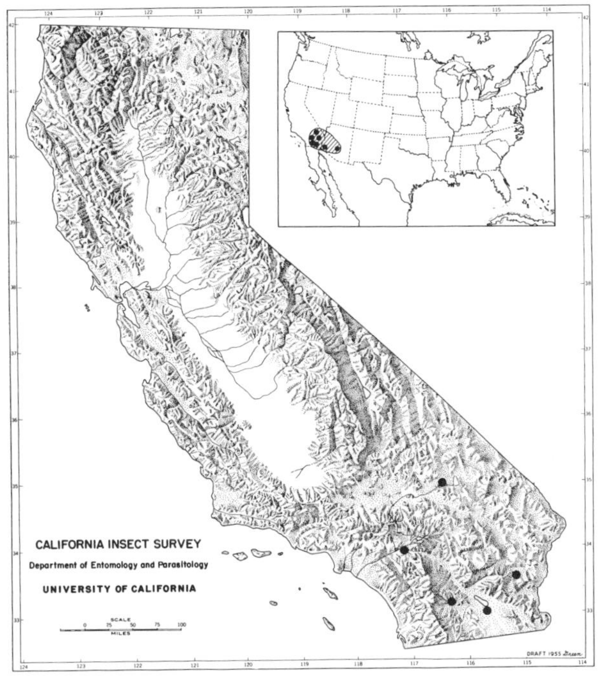

Map 23. California distribution of *Microstictia deserticola* Bohart. Inset: overall distribution.

Discussion. *Microstictia deserticola* differs from congeners in both sexes by the all- or nearly-all-yellow front and middle femora, short galea, and whitish clypeus. In addition, the male has separated projections on S-II (fig. 24n), unmodified femora, moderately long F-XI (2x breadth), narrow spiracular lobes, and an unusually broad sclerotized area on S-VII (fig. 24c). It is most similar to *lobata*, but *deserticola* is easily distinguished by the shorter galea (fig. 24i), and in the male by narrower spiracular lobes (fig. 24c), shorter F-XI, and separate projections on S-II.

Microstictia lobata Bohart, (new species) (Fig. 25; map 24)

Male holotype. Length 10 mm, forewing 7 mm. Black, with yellow as follows: pleuron mostly, propodeum including angles, legs except black marks on femora posteriorly, abdominal tergal bands laterally, S-I-III mostly, S-IV-VI in posterior bands. Whitish are: face (fig. 25d), including scape, pronotum, scutum laterally and wing base, discal scutal dots, scutellar band narrowed medially, metanotum, V-shaped mark on propodeal enclosure, broad tergal bands dorsally, three prongs of S-VIII, distal third of gonostylus, flagellum dull pale beneath, darkening beyond F-VI. Pubescence pale silvery appressed on clypeus, lower frons, and to some extent in many other areas. Punctures fine, inconspicuous. Facial proportions (fig. 25a, b, d, i, j): galea 2.3x as long as labrum, 1.25x eye height; palpi (fig. 25j); F-XI 2.5 times as long as broad, forebasitarsus 4x as long as broad, with 4 short rake setae (fig. 25g); femora and tarsi of midleg (fig. 25f) and hindleg unmodified; T-VII with spiracular lobes nearly half as broad as long (fig. 24c); S-II with a posteromedial projection which is notched apically (fig. 24n); S-VII (fig. 24c); S-VIII with a sharp medioventral tooth (fig. 25m); gonostylus (fig. 25k, L).

Female. Length 9-11 mm. Pale markings as in male, but more extensive and all yellow on notum and terga; scutum with 2 long discal dashes or an incomplete U; frons a little broader; forebasitarsus 2.6x as long as broad, with 4 well developed preapical rake setae; arolium as in Figure 24h.

Geographic range. Upper and Lower Sonoran localities in the Inyo Mts., San Bernardino to Imperial counties, California, and adjacent Yuma County, Arizona (map 24).

Records. Holotype male, Magnesia Canyon, Riverside Co., California, VI-24-52 (J. C. Hall, UCD). Paratypes, 6 males, 6 females. California: INYO CO.: Darwin Falls, V-21-65 (A. Menke); SAN BERNARDINO CO.: Cronise Valley, IV-29-56, on *Heliotropium curassavicum* (P. Hurd); Yermo, VII-12-57 (G. Beevor); RIVERSIDE CO.: Palm Springs, VI-12-54 (J. Hall); Magnesia Canyon, IV-29-56 and VII-2-52 (A. McClay, H. Mathis, E. Schlinger, J. Hall); Mecca, VIII-18-27 (J. Bradley); Ripley, IV-30-52 (G. Marsh); Happy Valley (T. Griswold); IMPERIAL CO.: Chocolate Mts., X-22-77, on *Hyptis emoryi* (M. Washbauer). Other paratypes, 1 male, 1 female. Arizona: YUMA CO.: 15 mi. n. Yuma, IX-3-57 (T. Haig); Yuma, V-6-39 (R. Bohart).

Discussion. *Microstictia lobata* has the tergal bands yellow (laterally only in male), a relatively long galea, and the clypeus whitish. Additional male characters are the unmodified femora, single median projection on S-II (fig. 25n), lengthened F-XI, broadest

Map 24. California distribution of *Microstictia lobata* Bohart. Inset: overall distribution.

spiracular lobes in the genus (fig. 25c), and strong medioventral tooth on S-VIII (fig. 25m). Females have black stripes posteriorly on the femora, which, in addition to the long tongue and narrower frons, distinguish *lobata* from *deserticola*.

Microstictia ocellata Bohart, (new species) (Fig. 26; map 25)

Holotype male. Length 9.5 mm, forewing 7 mm. Black, marked with yellow as follows: face mostly (fig. 26d); outer orbits, but broken at inner eye angle, pleuron mostly; legs except for large posterior stripes; T-I-III laterally; S-I-II mostly; others with broad bands. Whitish are: pronotal ridge and lobes, spot on tegula and adjacent scutum, triangular spots on scutellum, metanotum, edging on propodeal enclosure, subapical tergal bands dorsally, 3 prongs of S-VIII, distal half of gonostylus. Pubescence pale, appressed silvery on clypeus, lower frons, and mesopleuron. Punctation inconspicuous, scattered macropunctures on labrum basally. Facial proportions (fig. 26a, b, d): galea

Map 25. California distribution of *Microstictia ocellata* Bohart. Inset: overall distribution.

1.65x as long as labrum and 0.9x eye height; F-XI 2x as long as broad; forebasitarsus 5x as long as broad, with 4 short rake setae (fig. 26g); femora and tarsi of midleg (fig. 26f) and hindleg unmodified; T-VII with narrow spiracular lobes (fig. 26c); S-II with a well developed and separated pair of projections (fig. 26h); S-VII (fig. 26c); S-VIII with a long stout medioventral tooth (fig. 26m); gonostylus (fig. 26k, L).

Female. Length 9-10 mm. Markings as in male; frons a little broader; forebasitarsus 2.7x as long as broad, with 4 well developed preapical rake setae; arolia small, as in Figure 24h.

Geographic range. Upper Sonoran Life Zone in the Mojave Desert and Owens Valley, California, in lower mountain ranges of Nevada, and in Arizona and Utah (map 25).

Records. Holotype male, Weeks, Lyon Co., Nevada, VI-22-61 (F. D. Parker, UCD). Paratypes, 4 males, 9 females. Nevada: LYON CO.: Weeks, VI-22-61 (F. Parker), VI-27-73 (R. Bechtel, L. Barclay); Wilson Canyon, VI-27-73 (R. Bechtel, L. Barclay); LINCOLN CO.: Coyote Summit, VI-20-58 (R. Bechtel); 12 mi. nw. Coyote Summit, IX-31-68 (D. Miller, J. Lauck). Arizona: NAVAJO CO.: Show Low (M. Cazier, R. Schrammel). Utah: WASHINGTON CO.: Snow Canyon (G. Knowlton, W. Hanson); Ivins (G. Knowlton); Shivwits Indian Reservation (G. Knowlton, W. Hanson). Other paratypes, 3 males, 5 females. California: INYO CO.: Big Pine, VI-17-29 (E. Van Duzee); Antelope Springs, VI-15-61 (C. Toschi), VII-10-62 (A. Foster), VII-11-67 (R. Bohart); Inyo Mts., VI-1-57 (E. Van Dyke); SAN BERNARDINO CO.: Adelanto, VI-1-62 (M. Irwin).

Discussion. *Microstictia ocellata* is characterized in both sexes by the yellowish clypeus, whitish tergal bands which are yellowed laterally, and nearly all dark scutum. Also, the femora have rather large dark marks posteriorly. The male has additional characters: femora of midlegs and hindlegs unmodified, S-II with separated projections, T-VII spiracular lobes narrow and simple, and S-VIII with a large basomedial projection. Utah females have been taken on *Asclepias*.

Microstictia rufescens Gillaspy
(Map 26)

Microstictia rufescens Gillaspy 1983:284. Holotype male, Lajitas, Brewster Co., Texas (USNM).

Geographic range. Southern California, Arizona, New Mexico, southern Texas, Baja California Sur (La Paz), Nuevo Leon. The single record from Riverside County, California, is marked on Map 26.

California record. RIVERSIDE CO.: 1 male, Magnesia Canyon, VII-2-52 (D. E. Barcus, Gillaspy collection).

Discussion. The characteristic reddish tint bordering the whitish markings, and the extensively black clypeus, are absent or hardly developed in males from more western localities. Important structural features in the male are the relatively short galea (0.5x eye height); unmodified femora; S-II process single, but slightly bifurcate apically; F-XI short; T-VII spiracular lobes narrow; and S-VIII medioventral process blunt.

Flower records from out-of-state localities are *Acacia*, *Condalia*, *Lepidium*, *Tamarix* and *Colubrina*. Collection records are from April to August. We have seen a total of 35 males and 10 females.

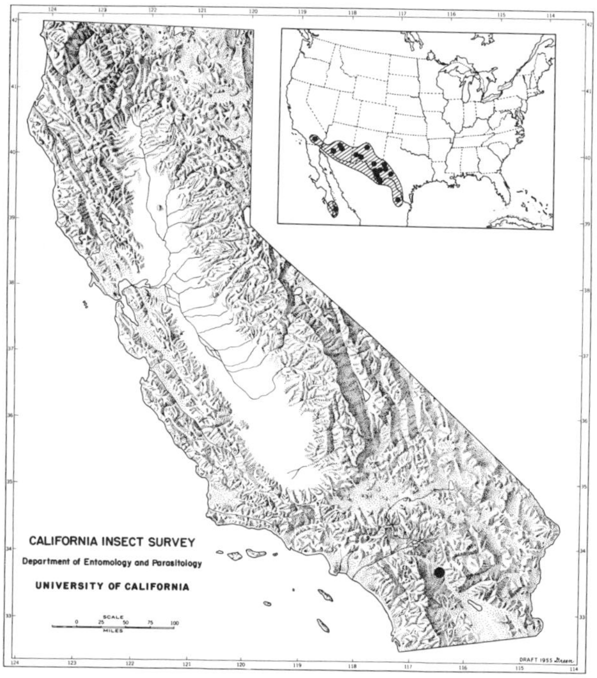

Map 26. California distribution of *Microstictia rufescens* Gillaspy. Inset: overall distribution.

Genus *Xerostictia* Gillaspy

Xerostictia Gillaspy 1963a:187. Generotype: *Xerostictia longilabris* Gillaspy, original designation.

This monotypic genus has been found only in desert areas of the southwestern United States and Baja California.

Biology. Alcock (1975) reported observations on a nesting site in the Gila River Indian Reservation east of Sierra Estrella, Maricopa County, Arizona. The wasps were nesting in a large stabilized sand dune during the hot part of July, and were associated with *Microbembex argyropleura* Bohart, *Glenostictia gilva*, and *G. argentata*. *Xerostictia* cells were 28-42 cm below the surface, and were stocked with adult *Brachynemurus longipalpis* (Neuroptera, Myrmeleontidae) and *Ormenis saucia* (Homoptera, Fulgoridae). This is the first record of antlion and planthopper prey use by a stictiellin wasp.

Burrows were closed while the female *Xerostictia* was away. Provisioning is progressive, as in species of *Steniolia*.

Taxonomy. Important morphological characteristics of *Xerostictia* are: (1) palpal formula 4-2; (2) labrum twice as long as broad (fig. 27d); (3) claws long and arolium small (fig. 27f); (4) midocellar plane a little longer than broad (fig. 27b); (5) clypeus not peaked above; (6) hindfemur with some very short erect hair beneath; (7) male midfemur serrate beneath (fig. 27j); (8) male midbasitarsus curved and with a group of setae in the concavity (fig. 27f); (9) male S-II edentate; (10) S-VI slightly incurved posteriorly; (11) male S-VIII with a spiniform medioventral projection (fig. 27m); (12) male T-VII ventral spiracular lobes short and posteriorly concavotruncate (fig. 27c); and male F-III-XI carinate and broadly fossulate beneath.

Characters 1, 2, and 7 above are of generic value; 3, 4, 5, 7, and 11 are perhaps of species-group significance; and the others may be largely specific. Except for the long labrum, reduced palpi, and moderately lengthened midocellus, *longilabris* would fit into *Stictiella*. Characters 5 and 7 are found in all *Stictiella*; 3 and 8 in most of them; and 6, 9, 10, and 11 in some of them.

Xerostictia longilabris Gillaspy
(Fig. 27; map 27)

Xerostictia longilabris Gillaspy 1963a:187. Holotype female, 12 mi. ssw. Ajo, Pima Co., Arizona (MCZ).
Xerostictia longilabris boharti Gillaspy 1963a:192. Holotype female, San Pedro, Baja California Sur (CAS). New synonymy.

Geographic range. Desert areas of s. California, sw. Arizona, Baja California, and Baja California Sur (map 27).

California records. 3 males, 6 females. RIVERSIDE CO.: Palms-to-Pines Highway at 3,200 ft., VII-27-50, on *Larrea diverticata* (P. Timberlake); Palm Springs, VI-16-75 (E. and J. Linsley); SAN DIEGO CO.: Borrego Valley, VIII-29-58 (E. Schlinger); IMPERIAL CO.: Ogilby Rd., 3 mi. s. junction with Highway 78, X-22-77, on *Hyptis emoryi* (M. Wasbauer); 3 mi. nw. Glamis, IX-16-72 (M. Wasbauer, A. Hardy).

Discussion. *Xerostictia longilabris* is remarkably similar to *Stictiella flavescens* and *S. villegasi* in size (14-16 mm long), robust shape, and markings. All three are extensively yellow with a prominent yellow U on

Map 27. California distribution of *Xerostictia longilabris* Gillaspy. Inset: overall distribution.

the scutum. However, there are many structural differences in *longilabris*, especially the long labrum, reduced palpi, and slightly lengthened midocellar plane (fig. 27b, d). These and other features are given above in the generic discussion. A few other points are: the moderately convex clypeus; the rather narrow frons, which is about equal to (female) or narrower than (male) the eye breadth (fig. 27d); and the galea longer than the compound eye. An additional male character is the slightly raised median longitudinal ridge on S-II.

Flower visitations are *Bebbia*, *Larrea*, *Lycium*, *Hyptis*, and *Wislizenia*.

Material from Baja California seems to be consistently a little different from Arizona and California collections. The former have more black on the terga and reduced scutal markings. Those who wish can assign the subspecies name of *boharti* to Baja California specimens.

Genus *Steniolia* Say

Bembex, *Monedula*, *Stictia* of authors.
Steniolia Say 1837:367, as a subgenus of *Bembex*. Generotype: *Bembex longirostra* Say 1837, monobasic.

The genus *Steniolia* contains 15 described species, all in North America, but 2 of them extend their range into n. South America (*guatemalensis* to Ecuador, and *longirostra* to Venezuela). In the United States and Canada, the ranges of the species with minor exceptions, are west of the 100th meridian. Four species (*guatemalensis*, *longirostra*, *powelli*, and *mexicana*) are not known to occur north of Mexico; one other essentially Mexican species (*dissimilis*) ranges into the mountains of southern Arizona, and 10 species occur in California.

Biology. Biological studies have been made of 8 species. A summary of known information was given by Evans and Gillaspy (1964) and Evans (1966). Nesting areas are localized and may contain fairly large aggregations. Adults visit flowers for nectar, and these long-tongued wasps can make use of flowers with deep corollas. Both sexes may spend the night on vegetation in ball-like clusters.

Females provision their nests with flies, particularly Bombyliidae and Syrphidae. Other dipterous prey noted in the literature are Sarcophagidae, Calliphoridae, Stratiomyidae, Muscidae, and Asilidae. The flies are paralyzed or often killed, and the wasp egg is laid on the first prey. Provisioning is fully progressive, and the burrow is temporarily closed while the wasp is away. Little is known about parasites, but cells have been found containing mutillid cocoons and those of the chrysidid *Parnopes edwardsii* (Cresson).

Taxonomy. The essential characters of *Steniolia* are: (1) labrum swollen toward base, blunt or emarginate at apex (fig. 28d); (2) palpal formula 3-2 or 3-1, midocellar plane slightly to strongly long oval; (3) arolium developed (fig. 29f); (4) tongue unusually long and reaching beyond hindcoxae when exserted; (5) clypeus weakly convex and in lateral profile nearly flat toward base; (6) hindfemur with abundant, erect hair beneath; (7) male S-VIII with a medioventral tooth, or rarely a raised angle (figs. 28m, 29m); (8) male S-II with a single posteromedian tooth or spine (fig. 28q).

Genus recognition in the field can be easily made by observing the swollen labrum shape with a hand lens. Also, the species are all fairly large, the body length generally 13-17 mm, thus differentiating *Steniolia* from smaller forms of related genera.

Gillaspy (1964) provided a detailed taxonomic revision of the genus, and the present treatment is a condensed and updated version of it, with emphasis on the California fauna. Gillaspy (1964) divided the genus into 3 species groups as follows:

1. The obliqua group, with flattened midtibia (fig. 29o), only slightly lengthened midocellar plane, curved male midbasitarsus (fig. 29f), and tyloides (specialized flattened areas) on male F-I-II (fig. 29h). Included species: obliqua, californiensis, and tibialis.
2. The sulfurea group, with characters of following group, but scape somewhat stouter, male spiracular lobes of T-VII unusually large (fig. 29c), and S-VIII of male unusually narrow at lateral prongs. Included species: sulfurea, nigripes, and dissimilis.
3. The longirostra group, with midtibia not unusually flattened (fig. 28o), midocellar plane long oval (fig. 28d), male midbasitarsus nearly straight (fig. 29f-eleg.), male F-I-II without tyloides, male T-VII spiracular lobes moderately broad to narrow (fig. 28c), male S-VIII not unusually narrow at lateral prongs. Included species: longirostra, guatemalensis, elegans, scolopacea, vanduzeei, duplicata, eremica, and mexicana.

KEY TO SPECIES OF STENIOLIA

1. Males, 7 visible terga, 11 flagellomeres 2

   Females, 6 visible terga, 10 flagellomeres 17

2. Midbasitarsus strongly curved (fig. 29f-tibi.) forebasitarsus with rake setae hardly evident, midtibia strongly flattened and expanded (fig. 29f-tibi.). 3

   Midbasitarsus practically straight (fig. 29f-eleg.), forebasitarsus usually with obvious rake setae, midtibia not unusually flattened or expanded (fig. 29f-eleg.). 5

3. Tergal markings whitish (w. U.S., British Columbia). . . . . . .obliqua (Cresson)

   Tergal markings yellow. 4

4. T-I with discal yellow spot or spots separated from lateral spots (as in fig. 30b), midtibia about 3x as long as broad (fig. 29f-tibi.) (Pacific Coast States). . . . . . .tibialis Handlirsch

   T-I with discal yellow area connected to lateral area (fig. 30a), midtibia ca. 3.5x as long as broad (s. California, Baja California). . . . . californiensis Gillaspy

5. Scape, tegula, posterior margins of several terga reddish 6

   Scape, tegula, tergal margins not reddish 8

6. Pale band of T-VI interrupted, body markings mostly whitish (Guatemala, s. Mexico, Ecuador) .guatemalensis Rohwer

   Pale band of T-VI not interrupted, body markings mostly yellow. 7

7. F-IX about 3x as long as broad, breadth of S-VIII at lateral projections in ventral view about half that at base of sternum (Arizona, n. Mexico) . . . . dissimilis C. Fox

   F-IX about 2.3x as long as broad, breadth of S-VIII at lateral projections in ventral view nearly as great as that at base of sternum (s. Mexico, Venezuela) . . . . longirostra Say

8. T-II-VI entirely yellow or almost so (fig. 30d); S-II apicomedial projection short, as in fig. 29m (California). . . . . sulfurea W. Fox

   T-II-VI with considerable black (fig. 30g, h), S-II apicomedial projection spinose (fig. 28m). 9

9. S-VIII with middle prong practically straight dorsally (fig. 29m), distitarsi partly or all black (fig. 29g) (e. and s. California, Baja California) . . . . nigripes J. Parker

   S-VIII with middle prong strongly curved (fig. 28m), tarsi various. 10

10. F-III and sometimes F-IV without a tyloid, T-II and T-III usually with a pair of isolated yellow spots (fig. 30h) (w. U.S., Mexico). . . . .elegans J. Parker

    F-III-IV with tyloides, T-II and/or T-III spots usually joined and connected to lateral yellow spots (fig. 30m). . . 11

11. F-IX with a ventral fossula extending along carina to distal fifth of flagellomere (fig. 28h), mesopleuron with anterior yellow band (or girdle) nearly always complete ventrally (sw. U.S., n. Mexico). . duplicata Provancher

F-IX with ventral fossula, when present, not extending into distal fifth of flagellomere (fig. 29h); mesopleural band (girdle) various. 12

12. F-IX with fossula absent or vestigial (fig. 29h), and carina appearing broadened into a long callous; notal and tergal markings nearly all whitish; mesopleuron with anterior yellow girdle often complete (e. California, w. Nevada). vanduzeei Gillaspy

F-IX with a distinct fossula (fig. 29h-scol.), markings various. 13

13. Mesopleural yellow girdle complete ventrally, scutum with submedial discal and posterior spots well developed, clypeus entirely yellow (sw. U.S., sw. Mexico). . . . . eremica Gillaspy

Mesopleural pale band (girdle) incomplete and often present only as a spot or two, clypeus often partly black (fig. 29s). 14

14. LID not more than two-fifths as great as head breadth, markings yellow (Mexico: Jalisco, Baja California) . . . . mexicana Gillaspy

LID more than two-fifths as great as head breadth, markings various. 15

15. Tergal markings (fig. 30k) yellow (California, north to Lassen Co.). . . . . scolopacea scolopacea Handlirsch

Tergal markings at least whitish medially 16

16. Labrum yellow, clypeus yellow and usually (90%) marked with black (nw. N. America including n. California and British Columbia) . . . . scolopacea albicantia J. Parker

Labrum yellow with black markings, clypeus whitish with black markings (Baja California). . .powelli Gillaspy

17. Scutum all dark, or with only a small pale spot adjacent to tegula. 18

Scutum with distinct parategular, discal, and often posteromedian pale spots. .22

18. S-II all yellow medially beneath overhanging basal flange (as in fig. 28p), midtibia slightly broadened (California). . . . . sulfurea W. Fox

S-II with black basally beneath overhanging basal flange, sometimes extensively black, midtibia various. . . .19

19. Midtibia only slightly broadened, S-II extensively black, clypeus and labrum partly black, tergal markings whitish (Baja California). . .powelli Gillaspy

Midtibia flattened and expanded to 2x its basal breadth or more (fig. 29o), clypeus and labrum not black-marked, other characters various. 20

20. Clypeus and terga with whitish markings (w. U.S., British Columbia) . . . . obliqua (Cresson)

Clypeus and terga with yellow markings 21

21. T-I with lateral yellow spots fused medially into a continuous band (as in male, fig. 30a) (s. California, Baja California). . californiensis Gillaspy

T-I with lateral yellow spots well separated (fig. 30b) (Pacific Coast states). . . . . . .tibialis Handlirsch

22. Tegula and at least some posterior tergal markings partly reddish. 23

Tegula and terga sometimes partly brown, but not distinctly reddish. 25

23. T-VI pale band partial, body markings mostly or all whitish (Guatemala, s. Mexico, Ecuador). . . . . .guatemalensis (Rohwer)

T-VI pale band complete, body markings yellow. 24

24. Femora with considerable reddish, scape about 3x as long as broad (s. Mexico, Venezuela). . . . . . longirostra (Say)

Femora with little or no reddish, scape distinctly less than 3x as long as broad (s. Arizona, Sonora). . . . . dissimilis C. Fox

25\. S-II all yellow medially beneath overhanging basal flange (fig. 28p). . . 26

S-II black medially beneath overhanging basal flange, black mark usually connected with basolateral black marks and sometimes with medial discal black (fig. 29r, p). 28

26\. Clypeus yellow medially but a little whitish around edges, scape about 2.9x as long as broad (sw. U.S., sw. Mexico). . . . . eremica Gillaspy

Clypeus whitish, scape various. 27

27\. Scape about 2.7x as long as broad, clypeus 2x as broad as long (e. and s. California, Baja California). . . . . .nigripes J. Parker

Scape about 2.9x as long as broad, clypeus 1.8x as broad as long (sw. U.S., n. Mexico). . . . .duplicata Provancher

28\. T-II with free or almost free posterodiscal yellow spots with axes converging anteriorly at about a right angle (fig. 30h), clypeus nearly always white (w. U.S., Mexico). . . . . elegans J. Parker

T-II with yellow discal spots, nearly always attached laterally, with axes converging anteriorly at an obtuse angle (fig. 30L, p), clypeal color various. 29

29\. Clypeus white, scutum with posteromedian yellow spot single and barlike (e. California, w. Nevada). .vanduzeei Gillaspy

Clypeus with at least some distinct yellow areas, scutum with posterodiscal yellow spot usually double, sometimes single and notched in front, rarely absent. 30

30\. Hindfemur without discal black anteriorly, S-II mostly yellow beneath overhanging basal flange (Jalisco, Baja California Sur). . . mexicana Gillaspy

Hindfemur with discal black anteriorly, S-II mostly or in large part black beneath overhanging basal flange (fig. 29p). 31

31\. California, n. to Lassen County (map 34) . . . . scolopacea scolopacea Handlirsch

Nw. N. America, including n. California, and British Columbia (map 34) . . . . scolopacea albicantia J. Parker

Steniolia californiensis Gillaspy
(Fig. 30a; map 28)

Steniolia californiensis Gillaspy 1964:52. Holotype male, Dollar Lake Trail, San Bernardino Mts., San Bernardino Co., California (ANSP).

Geographic range. Mountains of s. California from s. Kern County to Riverside County in Transition to Canadian life zones. S. californiensis has also been found in the Sierra San Pedro Martir of Baja California and near Loreto in Baja California Sur (map 28).

California records. KERN CO.: Frazier Park, Mill Potrero, Mt. Pinos; SAN BERNARDINO CO.: Barton Flats, Big Bear Lake, Camp Baldy, Onyx Peak, Mt. San Bernardino; LOS ANGELES CO.: Crystal Lake, Mt. Baldy Road, Mt. San Antonio; RIVERSIDE CO.: San Jacinto Mts.; SAN DIEGO CO.: Mt. Palomar.

Map 28. California distribution of Steniolia californiensis Gillaspy. Inset: overall distribution.

Discussion. The obliqua group, of which Steniolia californiensis is a member, is characterized especially by the flattened and expanded midtibia in both sexes. Other diagnostic features of the group are the distorted midbasitarsus in the male, and the extensive black markings at the base of S-II in the female. Other peculiarities are: the clump of small bristles anterodistally on the midfemur; the absence of discal or posteromedian pale spots on the scutum, and the near absence of rake setae on the male forebasitarsus. Further, the midocellar plane is only slightly lengthened, and the scape is rather stout. In males, F-I-II have tyloides ventrally, and S-VI is notched mediodistally.

Within its species group, californiensis differs from obliqua by the yellow rather than whitish notal and tergal markings, and from tibialis by the more continuous yellow of T-I (fig. 30a).

Flower records are Monardella and Eriodictyon. Specimens have been collected in each month from May to September. We have seen about 150 males and 80 females.

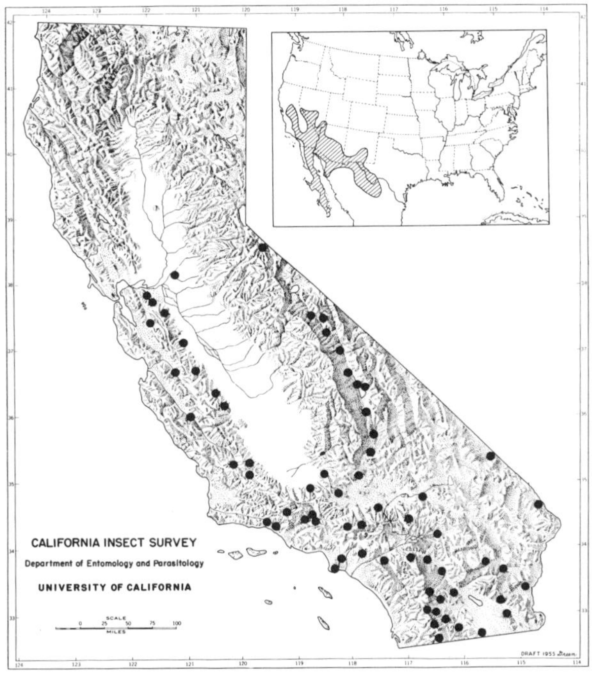

Map 29. California distribution of Steniolia duplicata Provancher. Inset: overall distribution.

Steniolia duplicata Provancher
(Figs. 28a-q, 30m-n; map 29)

Steniolia duplicata Provancher 1888:414. Lectotype male (Gahan and Rohwer 1918: 196), "Los Angeles" (Laval University, Quebec).

Steniolia meridionalis C. Fox 1923:430. Holotype male, Guaymas, Sonora, Mexico (CAS).

Geographic range. Mostly Upper and Lower Sonoran life zones in sw. U.S. from central California to s. Texas (1 male, Duval County), and in n. Mexico (map 29).

California records. Sacramento and Alameda counties to San Diego and Imperial counties, mostly in foothills and lower mountains (map 29).

Discussion. S. duplicata is one of the most abundant Steniolia in the southern U.S. west of the 100th meridian, and many hundreds of specimens have been examined. It is frequently taken along with scolopacea and elegans, both also relatively abundant. A separation of the three requires careful examination of key characters. Male duplicata have F-IX with a long ventral fossula; F-III carinate; and gonostylus apex almost always visible without dissection, and rounded, not acute as in elegans. Also, the clypeus is practically always without black markings, the mesopleuron is extensively yellow, and the medial pale spots of T-II are nearly always confluent with lateral spots. In females, S-II is all yellow basomedially (fig. 28p), the clypeus is whitish, the mesopleuron is extensively yellow, and the medial yellow spots of T-II are confluent with lateral yellow areas (fig. 30n).

A wide variety of flowers is visited, and the seasonal distribution is unusually long, late February to early December. Evans and Gillaspy (1964) presented detailed observations on clustering in Sacramento County, California, as well as in Brewster and Presidio counties, Texas. This phenomenon, which was first reported by Bradley (1908), was seen numerous times by Gillaspy during early morning and evening, or when weather was inclement. Both sexes were found in clusters. Nesting was observed by both Evans and Gillaspy in California and Texas. Burrows were in sandy earth and depth varied from 10 to 18 cm. The single cells were filled with bombyliids (Villa, Aphoebantus), syrphids (Paragus, Mesogramma, Eupeodes), sarcophagids (Sarcophaga), a muscid (Fannia), and calliphorids. The egg is apparently attached to the first fly prey, and provisioning is fully progressive. As in other Steniolia, burrows are temporarily closed during absence of the female wasp.

Steniolia elegans J. Parker
(Figs. 29f-eleg., 29r, 30h; map 30)

Steniolia elegans J. Parker 1929:50. Holotype male, San Luis Potosi, Mexico (USNM).

Geographic range. Widespread in Lower Sonoran to Canadian life zones of w. U.S. and Mexico (map 30). Nearly all records are east of the 100th meridian, but a few in Texas (Kingsville) and Mexico (Tamaulipas, Huisachal Canyon, 25 km sw. Ciudad Victoria) are closer to the 98th meridian.

California records. Widespread in the state (map 30).

Discussion. As the specific name implies, *elegans* is a beautifully marked species. The pattern of the abdominal dorsum is broken into 4 longitudinal rows of spots, the lateral ones triangular, the median ones long oval. The rows are nearly always separate on T-I-II; on T-III, at least in males, and tend to coalesce posterad. The free spots of T-II are distinctive in that their apparent axes converge anteriorly at nearly a right angle (fig. 30h), whereas the axes of attached spots or separated ones in related species form a broadly obtuse angle (fig. 30k, m). Additionally, males have the apicomedial process of S-VIII curved, midbasitarsus straight (fig. 30k, m), gonostylus apex almost always visible without dissection and acute laterally, and no tyloid on F-III. There is a tendency for the medial pairs of tergal spots of the male to be whitish. This condition is most marked in Mexican material. The female has S-II basomedially black (fig. 29r), as in *scolopacea* but not in *duplicata*. Further, the clypeus is nearly always white rather than partly yellow as in *scolopacea*.

Information on the biology has been given by Evans and Gillaspy (1964), and more extensively by Evans (1973). The larva was described by Evans and Lin (1956). A variety of flowers are visited by adults, which are on the wing from March to October. Adults are known to cluster as in other *Steniolia*. Nests are constructed in friable but stony soil. Burrows have a depth of 7-17 cm and contain a single cell. Prey reported by Evans (1973) are mostly Bombyliidae (*Poecilanthrax*, *Anthrax*, *Systoechus*, *Villa*), but also Asilidae (*Holopogon*) and Syrphidae (*Eupeodes*, *Microdon*, *Volucella*). The egg is laid on the first fly in the cell, and provisioning is fully progressive. When the nest is finally closed, the tumulus is not leveled.

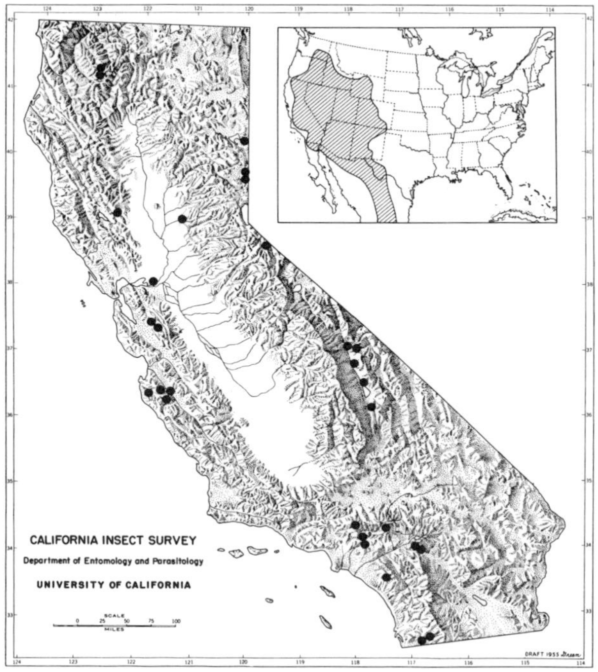

Map 30. California distribution of *Steniolia elegans* J. Parker. Inset: overall distribution.

Steniolia eremica Gillaspy
(Fig. 30i, j; map 31)

Steniolia eremica Gillaspy 1964:67. Holotype male, Calexico, Imperial Co., California (CAS).

Geographic range. Desert localities in Upper and Lower Sonoran life zones of s. California, Nevada, Utah, Arizona, Baja California, and Sonora (map 31).

California records. SAN BERNARDINO CO.: 17 mi. sw. Baker, Cronise Valley; RIVERSIDE CO.: 6 mi. w. Indio, Eagle Mt., Coachella, Thermal, Salton Sea; SAN DIEGO CO.: Warner Springs, Scissors Crossing; IMPERIAL CO.: Heber, Westmorland, San Felipe Creek, Harpers Well, Calexico.

Discussion. *Steniolia eremica* is related to *duplicata*, *elegans*, and *scolopacea*, but is much less abundant. In the male, the yellow of the mesopleuron is continuous ventrally, as in *duplicata* and most *elegans*. The short fossula beneath F-XI of *eremica* distinguishes from *duplicata*, and the continuous yellow markings across T-III and usually fused lateral and submedial spots of T-I (as in fig.

Map 31. California distribution of *Steniolia eremica* Gillaspy. Inset: overall distribution.

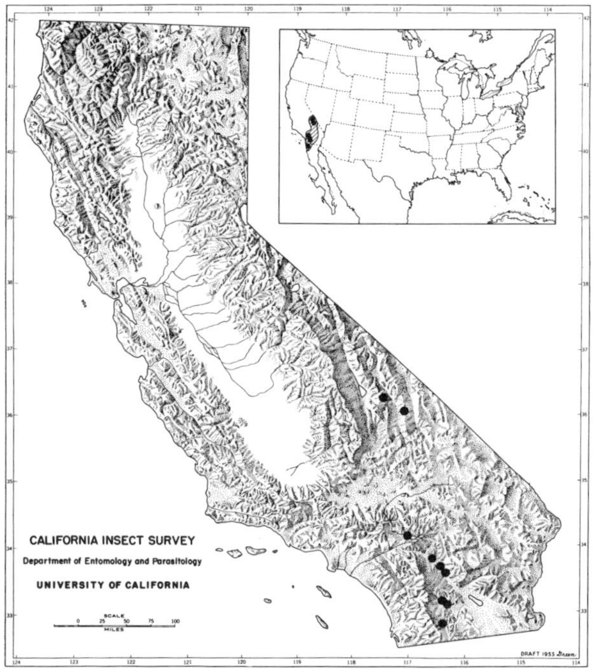

Map 32. California distribution of *Steniolia nigripes* J. Parker. Inset: overall distribution.

30i) are useful points of separation. *S. eremica* female has a shortened scape, and both it and *duplicata* have the clypeus whitish instead of mostly yellow.

Adults have been reported to visit flowers of *Palafoxia*, *Petalonyx*, *Melilotus*, and *Chilopsis*. They fly from March to August. Little is known of the biology. An asilid fly (*Efferia texana*) was pinned with a specimen from Riverside County and is presumably one of the prey (Evans and Gillaspy 1964). We have seen 20 males and 17 females.

Steniolia nigripes J. Parker
(Figs. 29c, g, m, 30f-g; map 32)

Steniolia nigripes J. Parker 1917:8. Holotype male, Los Angeles, Los Angeles Co., California (USNM).

Geographic range. Owens Valley and s. California in Upper and Lower Sonoran life zones (map 32). A record from Baja California is 6 mi. ne. Arroyo Seco.

California records. INYO CO.: Darwin Falls, Surprise Canyon; SAN BERNARDINO CO.: Banning, Berdoo Canyon, Palm Springs and nearby canyons, Palm Desert; SAN DIEGO CO.: Borrego Valley, Laguna Mts., Yaqui Well.

Discussion. The blackened distitarsi in the male (fig. 29g) are a striking feature on which the species name is based. However, this condition occurs at times in *scolopacea* and *duplicata*. The best single character for recognition is the middle prong of S-VIII, which is nearly straight dorsally and with a subapical tooth or "step" ventrally (fig. 29m). In the female, S-II is yellow basomedially and the clypeus is white, as in *duplicata*. The two species are not easy to separate, but *nigripes* has a slightly stouter scape (2.7 versus 2.9x as long as broad), and the clypeus is slightly broader and shorter (2x versus 1.8x).

The sole biological record was made by Gillaspy (1951), who observed a small nesting site at Yaqui Well, San Diego County, California. Cells were only 6-11 cm below ground and were stocked with Bombyliidae (*Aphoebantus*). Burrows were made in coarse sandy soil. The larva was described by Evans and Lin (1956). *S. nigripes* is essentially a vernal species, flying from March to June. We have seen about 70 males and 55 females.

Steniolia obliqua (Cresson)
(Fig. 30c; map 33)

Monedula obliqua Cresson 1865:469. Lectotype female (Cresson Jr. 1916:97), "Colorado" (ANSP).

Geographic range. W. U.S. and British Columbia in Canadian and Hudsonian life zone localities west of the 100th meridian. Californian localities are in the e. central part of the state (map 33).

California records. MONO CO.: White Mts. (10,000 ft.), VI-26 to VIII-26, 1960-1968 (various collectors); Fales Hot Springs, VIII-18-60 (D. Cavagnaro); Toms Place, VIII-8-62 (L. Stange, A. Menke); Pickel Meadow, VIII-11-60 (A. Menke); TUOLUMNE CO.: Chipmunk Flat, VIII-9-60 (M. Irwin); Tioga Pass; INYO CO.: 6 mi. w. Big Pine, IX-6-64 (J. Buckett); Mono Pass.

Discussion. As the nominal member of the *obliqua* group, *Steniolia obliqua* has the group characteristics listed under *californiensis*. The whitish notal and tergal markings are unique in the group. Other features are the moderately expanded midtibia, in contrast to *tibialis*; and customarily quadripartite pale markings of T-I (fig. 30c), rather than at most bipartite as in *californiensis*.

Evans (1964) has given features of the larva. The biology was discussed by Evans and Gillaspy (1964) and Evans (1966, 1970).

Adults frequent flowers of various kinds, and the season extends from June to September. During inclement weather or overnight, clusters consisting of many individuals of both sexes sometimes form. Females seem to be at the center of such clumps, with males on the outside. Mating may occur in clusters or in nesting sites. Nests are made in finegrained sandy loam rather than in pure sand. Recorded prey have been mostly Bombyliidae: *Bombylius*, *Villa*, and *Geron*. Other records are *Pipiza* (Syrphidae) and *Lucilia* (Calliphoridae). The egg is glued erect to the side of the first fly prey in the cell. Provisioning is progressive, with temporary closures until the larva is mature. The cocoon is ovoid and surrounded by a shroud-like envelope which bears fragments of prey and has a broadly funnel-shaped structure at one end (Evans 1966:461). The chrysidid parasite *Parnopes edwardsii* (Cresson) was reared by G. E. Bohart from an *obliqua* cocoon in Idaho. Also, he found *mutillids*, probably *Dasymutilla*, attacking cocoons (Evans 1966:219).

Map 33. California distribution of *Steniolia obliqua* (Cresson). Inset: overall distribution.

The flight season is June to September. We have seen about 160 males and 140 females.

Steniolia scolopacea scolopacea Handlirsch
(Figs. 29h-scol., 29p, s, 30k-L; map 34)

Steniolia scolopacea Handlirsch 1889:510. Lectotype female (Gillaspy 1964:71), "Nevada" (Vienna Mus.).

Geographic range. Upper Sonoran to Canadian life zones in California, e.-central Nevada, and Baja California Sur. The typical subspecies is replaced by the whitish-marked *scolopacea albicantia* in ne. California, Oregon, Washington, Idaho, and British Columbia (map 34). Both yellow and whitish males may be found in Modoc, Siskiyou, and Lassen counties of California. There are no records of *scolopacea* from se. California.

California records. Most counties from Lassen to Kern, and Coast Range localities from Monterey to San Diego counties (map 34).

Discussion. Three related species are relatively abundant in California, scolopacea, elegans, and duplicata, and their distinction is not always an easy matter. In male scolopacea, the clypeus is often black-marked (fig. 29s) and not so in the others. Further, the fossula beneath F-IX does not extend into the distal fifth (fig. 29h-scol.), and the submedial pale spots of T-II are not always connected to lateral spots (fig. 30k). Female scolopacea have a black margin at the base of S-II (fig. 29p), T-II spots as in most males (fig. 30L), and the clypeus is at least partly yellow.

A variety of flowers are visited during the flight season of May to October. We have seen about 600 males and 500 females.

Steniolia scolopacea albicantia J. Parker (Map 34)

Steniolia albicantia J. Parker 1917:12. Holotype, Columbia River, Grand Coulee, Washington (USNM). Holotype data amplified by Cockerell (1918).

Geographic range. Nw. U.S. and British Columbia in Upper Sonoran and Transition life zones (map 34). California localities are in the ne. corner of the state and as far south as Hallelujah Junction, Lassen County.

California records. MODOC CO.: Cedarville, Lava Beds National Monument, Davis Creek, Hackamore, Lake City; LASSEN CO.: Hallelujah Junction.

Discussion. No reliable structural differences have been found in the 2 subspecies of scolopacea. Also, females are separable only on the basis of locality. In males, the whitish color, especially on the more medial markings of the terga, and sometimes on the notum, is striking. Furthermore, there is a strong geographical correlation, with Oregon, Washington, and Idaho males 100% whitish, ne. California males 70%, and s. Lassen County males nearly 50%. For this reason we have maintained the subspecies status. Albinism occurs in other stictiellins also, but where the geographical separation is doubtful, such as in Stictiella emarginata, and even in Steniolia elegans, we have not used the subspecies concept. A discussion of relationships between scolopacea and albicantia has been given by Gillaspy (1964:77).

We have seen about 160 males and 70 females collected from June to October.

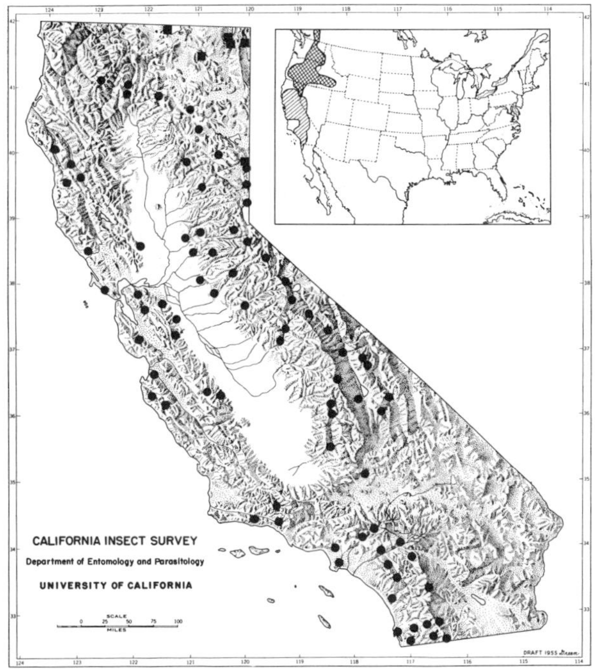

Map 34. California distribution of Steniolia scolopacea scolopacea Handlirsch (black circles) and S. scolopacea albicantia J. Parker (black squares). Inset: overall distribution. Crosshatched part of figure refers to albicantia.

Steniolia sulfurea W. Fox (Figs. 29h-sulf., 29q, 30d-e; map 35)

Steniolia sulfurea W. Fox 1901:84. Lectotype male (Gillaspy 1964:105), "Shasta County, California" (AMNH).

Geographic range. Upper Sonoran Life Zone in California, from Shasta and Humboldt counties south to Fresno County (map 35).

California records. 36 males, 26 females, June 12 to August 6. SHASTA CO.: no detail; HUMBOLDT CO.: 2 mi. e. Garberville; MENDOCINO CO.: 4 mi. w. Eel River Ranger Station; STANISLAUS CO.: Del Puerto Canyon and 5 mi. w.; SANTA CLARA CO.: Colorado Creek (29 mi. s. Livermore), San Antonio Valley, Mt. Hamilton; MARIPOSA CO.: no detail; FRESNO CO.: Squaw Valley, Tollhouse.

Discussion. Steniolia sulfurea is a rare and strikingly marked species, the thorax mostly dark, legs mostly yellow, and abdomen nearly all yellow (male, fig. 30d), or extensively so (female, fig. 30e). In addition to its outstanding color pattern, *sulfurea* has several other peculiarities. It belongs to the *sulfurea* species group, which has the scape unusually stout, midocellar plane long oval, midtibia not unusually flat, male midbasitarsus straight, male T-VII spiracular lobes quite large, and male S-VIII rather narrow. The male of *sulfurea* differs from *nigripes* by the scutum all dark, abdomen and legs nearly all yellow, S-II dentate but not spinose posteriorly (fig. 29q), T-VII spiracular lobes mostly white instead of mostly brown, and S-VIII middle prong more curved and more strongly toothed ventrally. Females of both species have the clypeus whitish and S-II yellow basomedially. However, *sulfurea* has the scutum and propodeum posteriorly all black or practically so, and the mesopleuron with restricted yellow spots.

Little is known of the biology. R. M. Bohart and L. S. Kimsey discovered a nesting site (VI-12-79) in the low mountains of w. Stanislaus County, California. A few male *sulfurea* were present, and perhaps a dozen females were constructing burrows in a gravel-sand area with sparse grass surrounded by *Adenostoma* bushes.

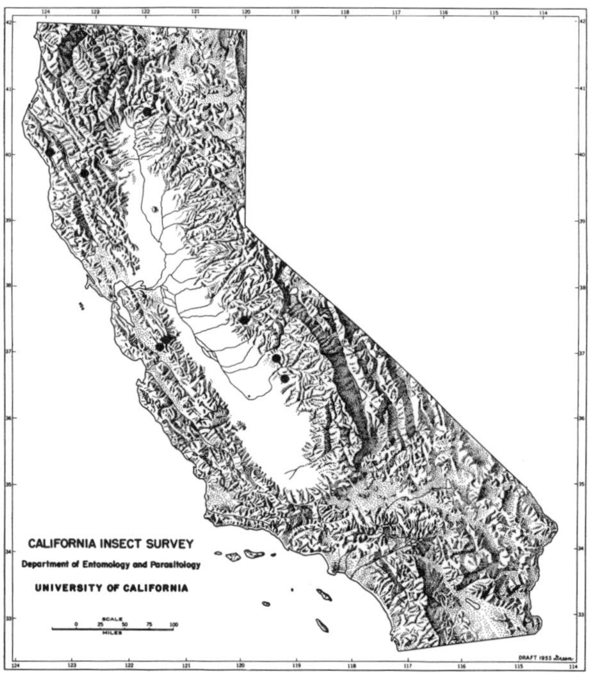

Map 35. California distribution of *Steniolia sulfurea* W. Fox.

Steniolia tibialis Handlirsch
(Figs. 29f-tibi., 29o, 30b; map 36)

Steniolia tibialis Handlirsch 1889:513. Lectotype male (Gillaspy 1964:56), "Nevada" (Vienna Mus.).

Geographic range. Mostly Canadian Life Zone localities in the Sierra Nevada of California, and in Oregon, Idaho, and Washington (map 36).

Discussion. Characters of the *obliqua* group, to which *Steniolia tibialis* belongs, are listed under *californiensis*. From its close relatives, *tibialis* differs by its broader midtibia (fig. 29o, f-tibi.), by the yellow markings (from *obliqua*), and from *californiensis* by separation of the medial and lateral pale spots of T-I (fig. 30b).

Adults visit a variety of flowers such as *Cirsium*, *Monardella*, and *Rudbeckia*, and the flight period is June to October. Prey recorded by Evans and Gillaspy (1964) were bombyliids (*Conophorus*) and syrphids (*Syrphus*). *S. tibialis* is abundant within its montane habitat, and we have seen about 400 males and 300 females.

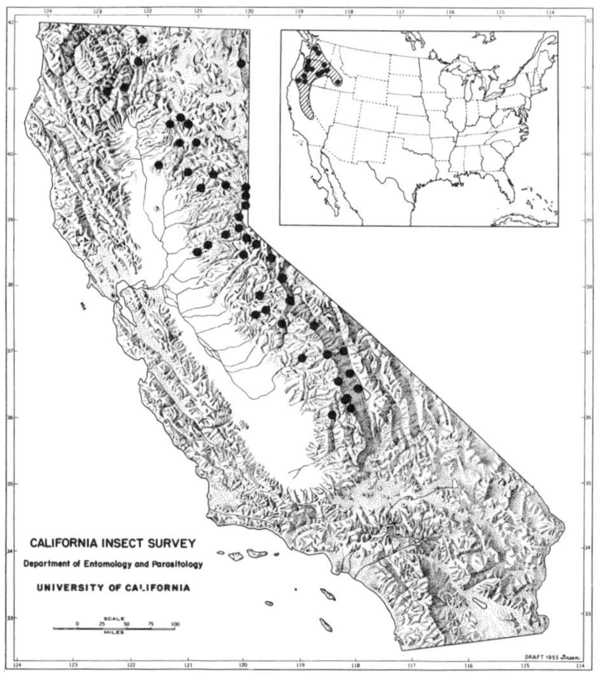

Map 36. California distribution of *Steniolia tibialis* Handlirsch. Inset: overall distribution.

Steniolia vanduzeei Gillaspy
(Figs. 29h-vand., 30o, p; map 37)

Steniolia vanduzeei Gillaspy 1964:62. Holotype male, Nixon, Washoe Co., Nevada (CAS).

Geographic range. Upper Sonoran and Transition life zones in e.-central California and w.-central Nevada (map 37).

California records. LASSEN CO.: Hallelujah Junction, VII-1-74 (R. Bohart), for other records from this locality see Gillaspy 1964; 7 mi. n. Hallelujah Junction, VIII-2-51, on *Cirsium* (J. Gillaspy, D. MacNeill); Herlong, IX-4-69 (R. Allen); MONO CO.: Mammoth, VIII-1-36 (G. and R. Bohart); Bridgeport, XI-10-57, on *Chrysothamnus* (J. Gillaspy); INYO CO.: Owens Valley, VIII-2-36 (G. and R. Bohart).

Discussion. As a member of the *longirostra* species group, *Steniolia vanduzeei* has the midtibia not unusually expanded, the male midbasitarsus straight, and the male S-VIII not particularly narrowed. Males of *vanduzeei* have notal and tergal markings whitish, as in *scolopacea albicantia*. However, *vanduzeei* has flattened tyloides beneath F-IX-X instead of well-marked fossulae (fig. 29h-vand.). Further, the male clypeus is rather whitish but not black-marked, in contrast to many *albicantia*. Females of the two forms are similar, too. In *vanduzeei* the clypeus is entirely white, whereas in *albicantia* it is tinged with yellow, admittedly a subtle distinction.

Nothing has been reported on biology. The flight period is June to September, and flowers visited are *Chrysothamnus*, *Cirsium* and *Tetradymia*. *S. vanduzeei* is an uncommon species, and we have seen only about 15 males and 30 females.

Map 37. California distribution of *Steniolia vanduzeei* Gillaspy.

Literature Cited

Alcock, J.
1975. The behavior of some bembicine wasps of southern Arizona. Southwestern Nat. 20:337-342.

Alcock, J., and G. J. Gamboa
1975. The nesting behavior of some sphecid wasps of Arizona, including Bembix, Microbembex, and Philanthus. Ariz. Acad. Sci. 10: 160-165.

Bohart, R. M.
1970. New species, synonymy, and lectotype designations in North American Bembicini. Pan-Pac. Ent. 46: 201-207.

1982. New species of Stictiella. J. Kansas Ent. Soc. 55:593-597.

1983. New species of Glenostictia. Proc. Ent. Soc. Wash. 85:485-492.

Bohart, R. M., and D. S. Horning
1971. California bembicine sand wasps. Bull. Calif. Insect Surv. 13: 1-49.

Bohart, R. M., and L. S. Kimsey
1982. A synopsis of the Chrysididae in America north of Mexico. Mem. Amer. Ent. Inst. 33:1-266.

Bohart, R. M., and A. S. Menke
1976. Sphecid wasps of the world: A generic revision. ix + 695 pp. Univ. of California Press, Berkeley.

Bradley, J. C.
1908. A case of gregarious sleeping habits among aculeate Hymenoptera. Ann. Ent. Soc. Amer. 1:127.

Cockerell, T. D. A.
1918. The bembicine wasps. Ent. News 29:59-60.

Cresson, E. T.
1865. Catalogue of Hymenoptera in the collection of the Entomological Society of Philadelphia, from the Colorado Territory. Proc. Ent. Soc. Phila. 4:242-313, 426-488.

Cresson, E. T., Jr.
1916. The Cresson types of Hymenoptera. Mem. Amer. Ent. Soc. 1:1-141.

1928. The types of Hymenoptera in the Academy of Natural Sciences of Philadelphia, other than those of Ezra T. Cresson. Mem. Amer. Ent. Soc. 5:1-90.

Evans, H. E.
1964. Further studies on the larvae of digger wasps. Trans. Amer. Ent. Soc. 90:235-299.

1966. The comparative ethology and evolution of the sand wasps. xvi + 526 pp. Harvard Univ. Press, Cambridge.

1970. Ecological-behavioral studies of the wasps of Jackson Hole, Wyoming. Bull. Mus. Comp. Zool. 140: 451-509.

1973. Notes on the nesting behavior of Steniolia elegans. Great Basin Nat. 33:29-30.

Evans, H. E., and J. E. Gillaspy
1964. Observations on the ethology of digger wasps of the genus Steniolia. Amer. Midland Nat. 72:257-280.

Evans, H. E., and C. S. Lin
1956. Studies on the larvae of digger wasps. Part II: Nyssoninae. Trans Amer. Ent. Soc. 82:35-66.

Fox, C. L.
1923. Expedition of the California Academy of Sciences to the Gulf of California in 1921: The Bembicini. Proc. Calif. Acad. Sci. (ser. 4) 12:429-436.

Fox, W. J.
1895. Synopsis of the Bembicini of boreal America. Proc. Acad. Sci. Phila. 1895:351-374.

1901. Two new bembicine wasps. J. New York Ent. Soc. 9:84.

Gahan, A. B., and S. A. Rohwer
1918. Lectotypes of the species of Hymenoptera (except Apoidea) described by Abbé Provancher. Canad. Ent. 56:196-201.

Gillaspy, J. E.
1951. Nesting habits of *Steniolia nigripes* Parker. Pan-Pac. Ent. 27:167-168.

1959. A new bembicine wasp related to *Stictiella tenuicornis* (Fox), with certain phylogenetic considerations. Pan-Pac. Ent. 35: 187-194.

1963a. Two new genera and a new species of Bembicini from North America, with a key to the genera having recessed ocelli. Ent. News 74: 187-199.

1963b. The identity of *Stictiella corniculata* Mickel, with a note on synonymy in *Stictiella*. Ent. News 74:250-252.

1964. A revisionary study of the genus *Steniolia*. Trans. Amer. Ent. Soc. 89:1-117.

1983. New taxa of bembicine Sphecidae. Southwestern Nat. 8:279-291.

Gillaspy, J. E., H. E. Evans, and C. S. Lin
1962. Observations on the behavior of digger wasps of the genus *Stictiella*, with a partition of the genus. Ann. Ent. Soc. Amer. 55: 559-566.

Handlirsch, A.
1889. Monographie der mit *Nysson* und *Bembex* verwandten Grabwespen: IV. Sitzunber. Akad. Wiss. Wien 98:440-517.

1890. Monographie der mit *Nysson* und *Bembex* verwandten Grabwespen: V. *Monedula* Sitzunber. Akad. Wiss. Wien 99:77-166.

Krombein, K. V.
1964. Miscellaneous prey records of solitary wasps: V. Bull. Brooklyn Ent. Soc. 58:118-120.

1979. Subfamily Bembicinae. *In* K. V. Krombein, P. D. Hurd, D. R. Smith, and B. D. Burks, Catalog of Hymenoptera in America north of Mexico, vol. 2, xiv + pp. 1199-2209. Smithsonian Press, Washington, D.C.

LaRivers, I.
1942. Notes on the bembicid, *Stictiella pulla* (Handlirsch). Pan-Pac. Ent. 18:4-8.

Mickel, C. E.
1918. New species of Sphecoidea from the central and western states. Univ. Studies, Lincoln, Nebr., 17:319-341 (1917).

Parker, J. B.
1917. A revision of the bembicine wasps of America north of Mexico. Proc. U.S. Natl. Mus. 51:1-155.

1929. A generic revision of the fossorial wasps of the tribes Stizini and Bembicini. Proc. U.S. Natl. Mus. 75:1-230.

Provancher, L.
1888. Additions et corrections au volume II de la Faune Entomologique du Canada traitant des Hymenopteres. Darveau, Quebec.

Say, T.
1837. Descriptions of new North American Hymenoptera and observations on some already described. Boston J. Nat. Hist. 1:361-416.

Smith, F.
1856. Catalogue of the hymenopterous insects in the collection of the British Museum: Part IV Sphecidae, Larridae, and Crabronidae, pp. 207-497. London.

Index

INDEX TO GENERA, SPECIES, AND SUBSPECIES OF STICTIELLINA[1]

[1] Genera are capitalized; synonyms are underlined.

Plates

Explanation of Illustrations

Drawings are not to scale. The midocellar area is alike in the two sexes, so only the male is shown. This area is bounded above and laterally by a groove, which is shown by a solid line. Not all pubescence is indicated, as on facial figures. Pigmentation of the flagellum is not shown except in figs. 2a, 3a, 10a, 12a, and 20a.

PLATE I

Fig. 1. Stictiellina dendrogram. (See text under Phylogeny for explanation of symbols.)

Fig. 2. *Glenostictia angulata* Gillaspy

a. flagellum, male.

b. pattern of scutum, scutellum, metanotum.

c. S-VII with overlapping (lateral) spiracular lobes of T-VII.

d. face, male.

e. tergal pattern, male.

f. S-II, male, oblique lateral view.

g. gonostylus, dorsal.

h. gonostylus, lateral.

i. S-VIII, male, lateral.

j. midocellar area, male.

k. S-VI, male.

l. midtarsus, male, lateral.

m. foretarsus, male, outer view.

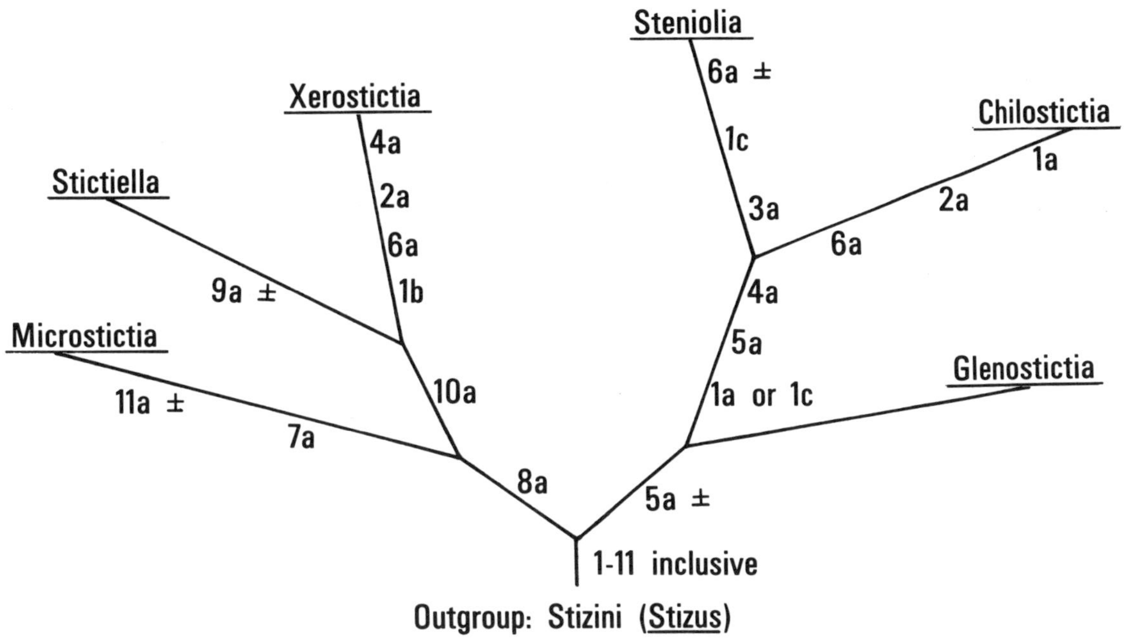

1. Dendrogram

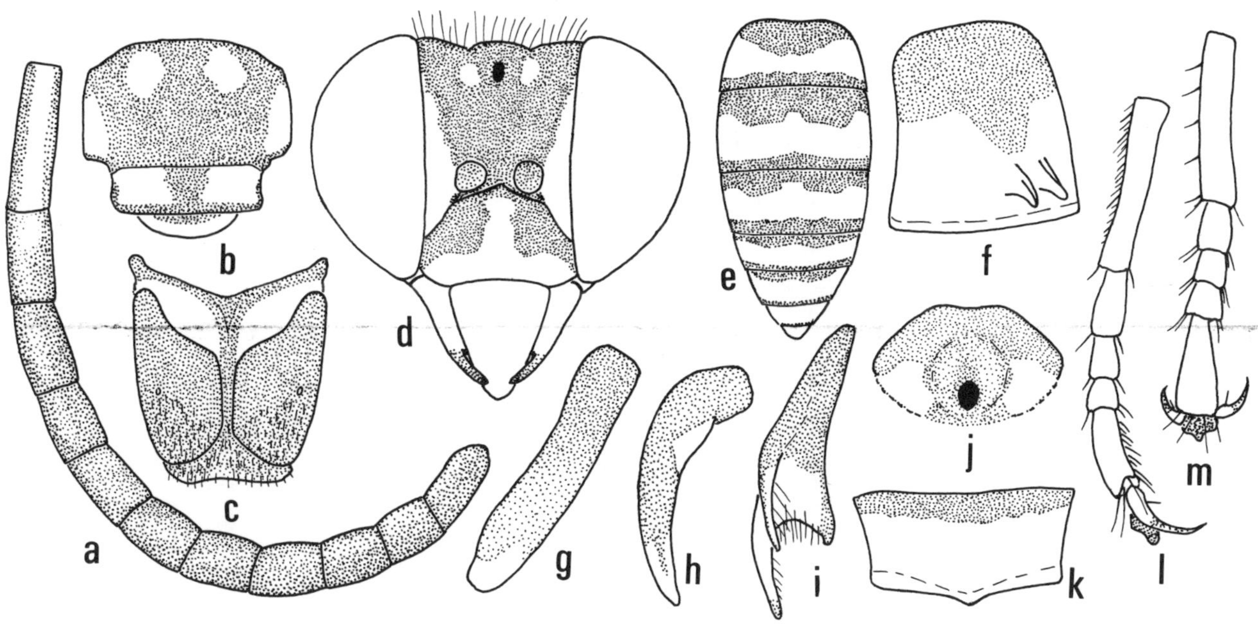

2. angulata

PLATE II

Figs. 3 and 4. *Glenostictia angulifera* Bohart and *Glenostictia argentata* (C. Fox).

a. flagellum, ventral, male.

b. midocellar area, male.

c. S-VII and overlapping lobes of T-VII, male.

d. face, male.

e. abdominal pattern, dorsal, male.

f. midtarsus, lateral, male.

g. foretarsus, dorsal, male.

h. S-II, oblique lateral, male.

i. gonostylus, dorsal.

j. gonostylus, lateral.

k. S-VIII, lateral, male.

l. forebasitarsus, dorsal, female.

m. S-VI, male.

n. T-II pattern, female.

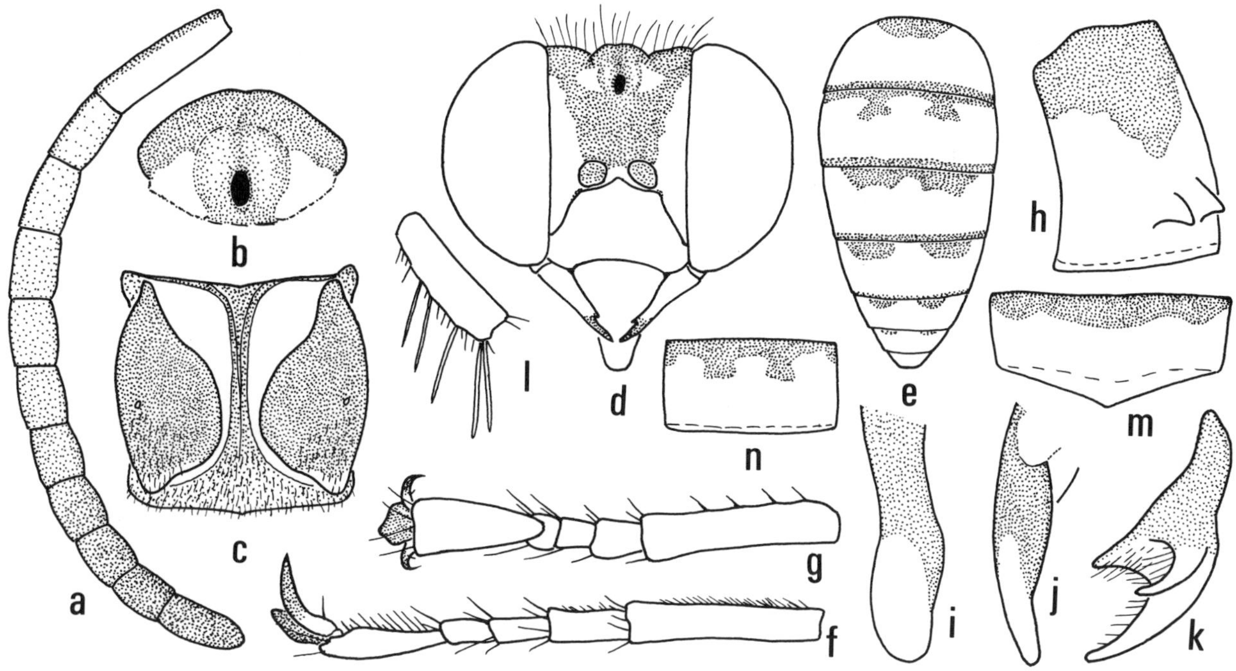

3. angulifera

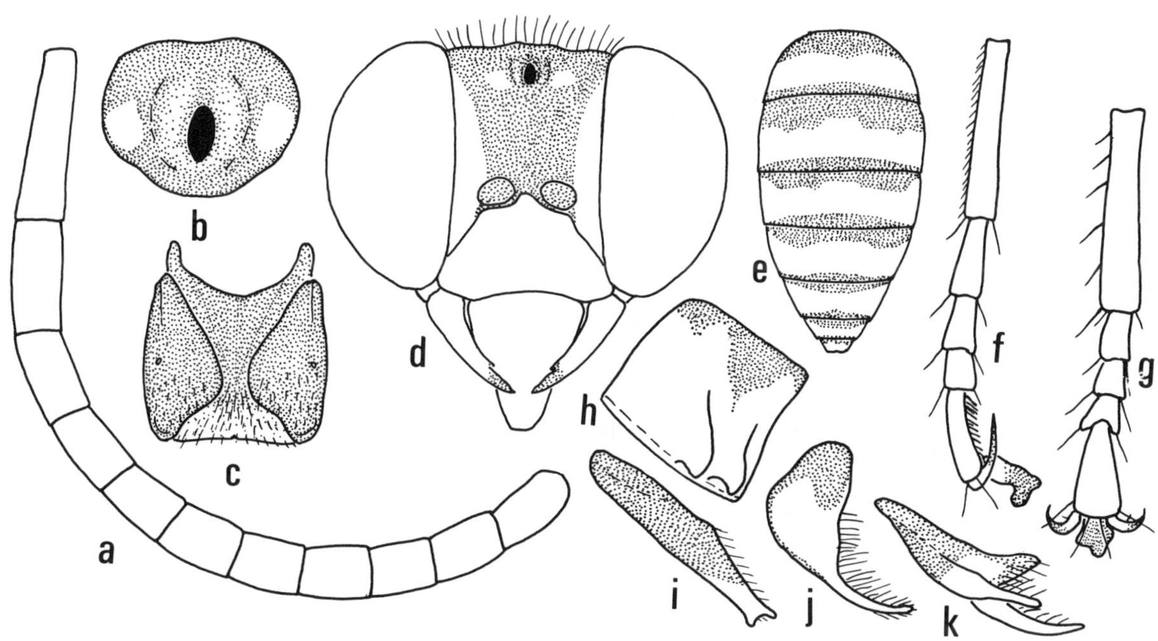

4. argentata

PLATE III

Figs. 5 and 6. *Glenostictia arizonae* Bohart and *Glenostictia bifurcata* (C. Fox).

a. antenna, ventral, male.

a*. flagellum, ventral, male.

b. midocellar area, male.

c. S-VII and overlapping lobes of T-VII, male.

d. face, male.

e. abdominal pattern, dorsal, male.

f. foretarsus, dorsal, male.

g. midtarsus, lateral, male.

h. base of S-II, female.

i. S-II pattern, male.

j. foretarsomeres II-IV, dorsal, male.

k. gonostylus, dorsal.

l. gonostylus, lateral.

m. S-VIII, lateral, male.

n. S-II, oblique lateral, male.

o. hindfemur, outer lateral, male.

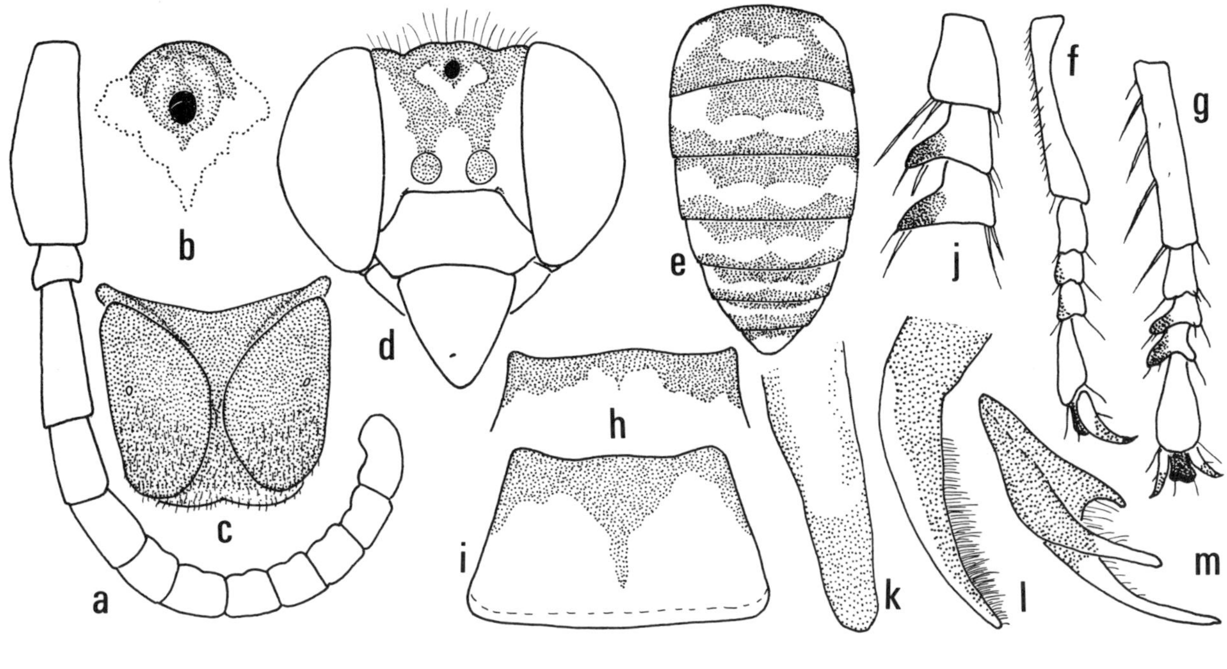

5. arizonae

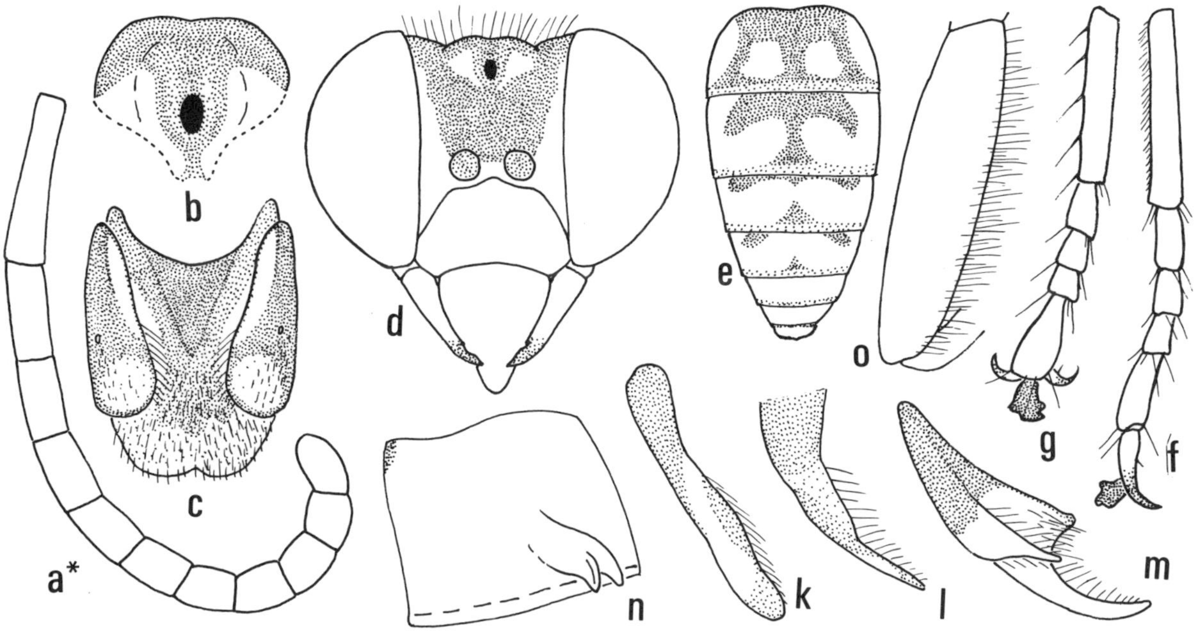

6. bifurcata

PLATE IV

Figs. 7 and 8. *Glenostictia bituberculata* (J. Parker) and *californica* Bohart.

a. flagellum, ventral, male.

b. midocellar area, male.

c. S-VII and overlapping lobes of T-VII, male.

d. face, male.

e. abdominal pattern, dorsal, male.

f. midtarsus, lateral, male.

g. foretarsus, dorsal, male.

h. abdominal pattern, dorsal, female.

i. clypeus, female.

j. S-II, oblique lateral, male.

k. gonostylus, dorsal.

l. gonostylus, lateral.

m. S-VIII, lateral, male.

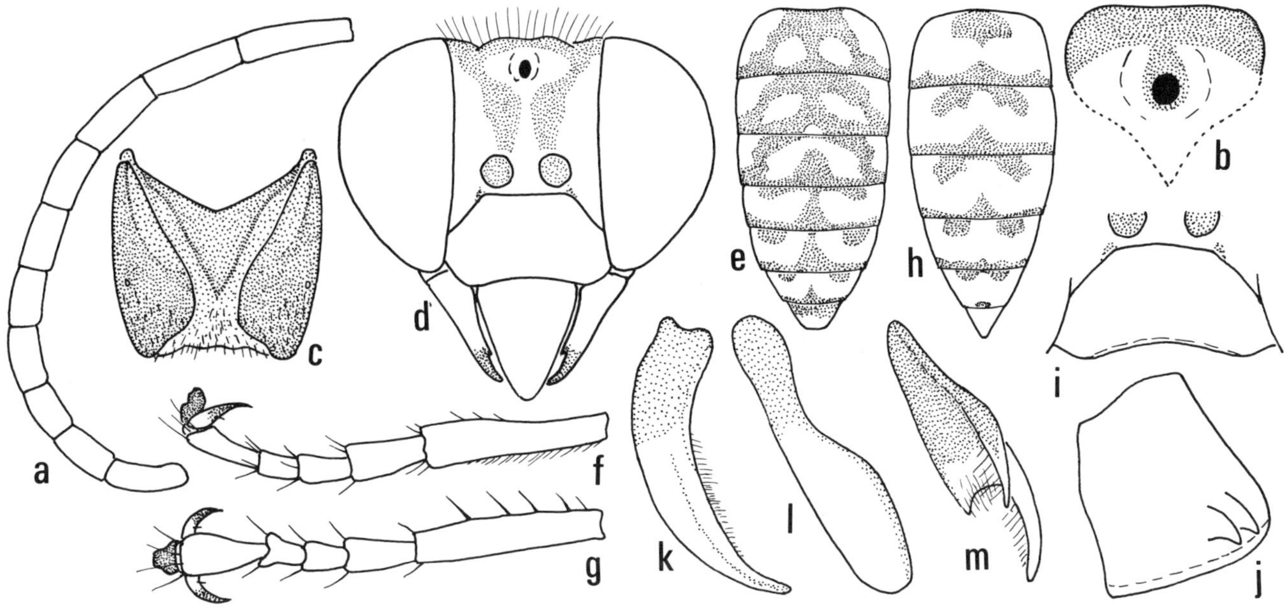

7. bituberculata

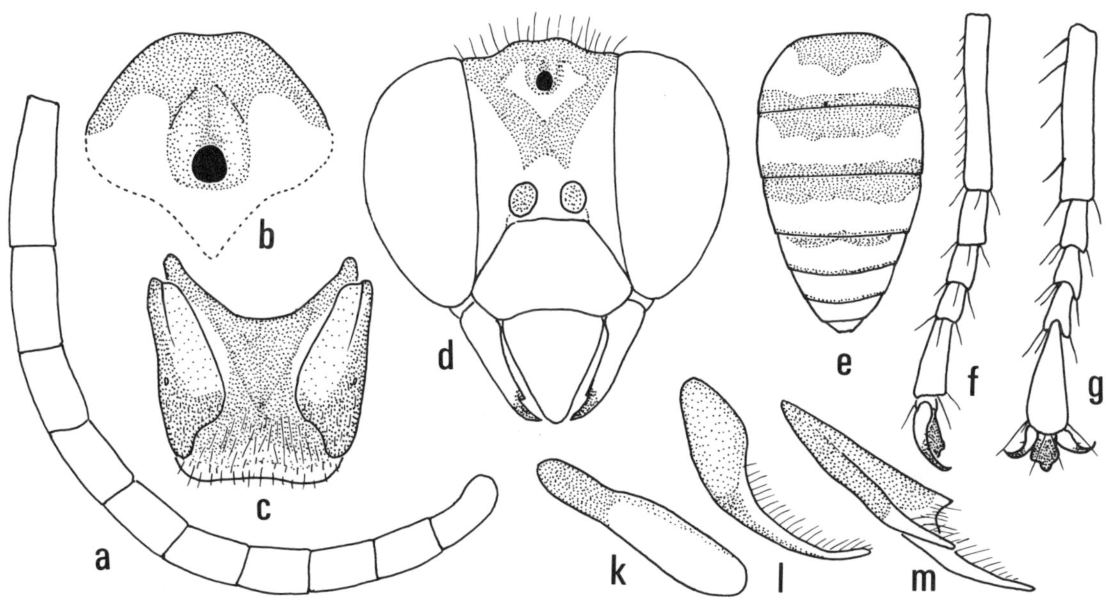

8. californica

PLATE V

Figs. 9 and 10. *Glenostictia clypeata* (Gillaspy) and *Glenostictia gilva gilva* Gillaspy.

a. flagellum, ventral male.

b. midocellar area, male.

c. S-VII and overlapping lobes of T-VII, male.

d. face, male.

e. abdominal pattern, dorsal, male.

f. midtarsus, lateral, male.

f*. foretarsus, lateral, male.

g. foretarsus, dorsal, male.

h. forebasitarsus, dorsal, female.

i. clypeus, female.

j. S-II base pattern, female.

k. gonostylus, dorsal.

l. gonostylus, lateral.

m. S-VIII, lateral, male.

n. mesonotal and metanotal pattern, male, *gilva gilva*.

o. mesonotal and metanotal pattern, female, *gilva gilva*.

p. S-VI, male.

q. S-II oblique lateral, male.

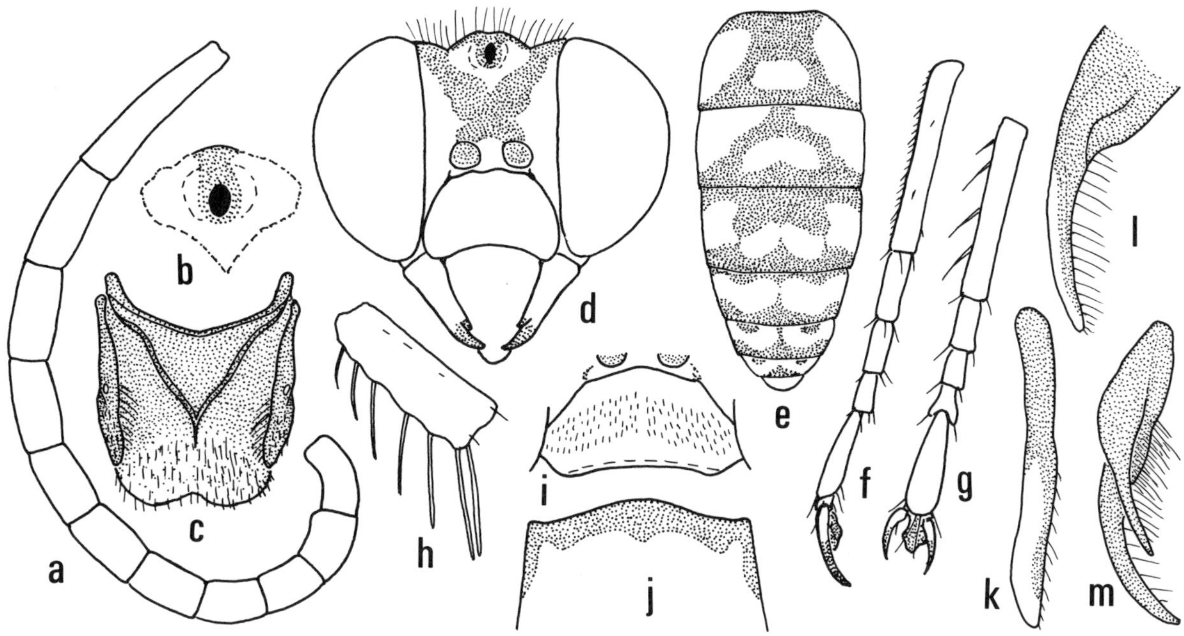

9. clypeata

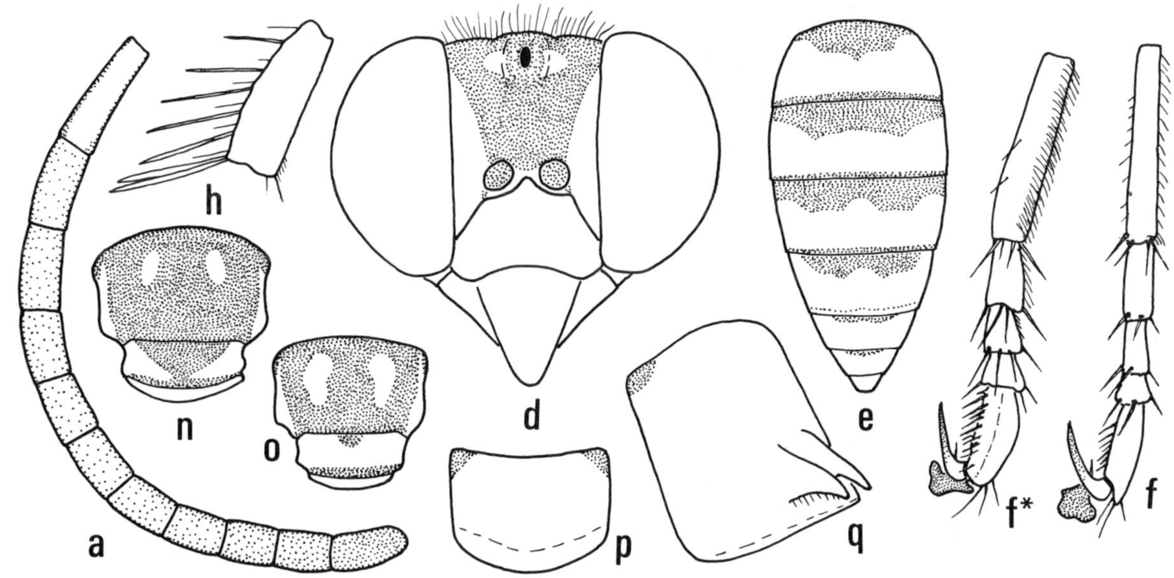

10. gilva

PLATE VI

Figs. 11 and 12. Glenostictia megacera (J. Parker) and Glenostictia pulla (Handlirsch).

a. flagellum, ventral, male.

b. midocellar area, male.

c. S-VIII and overlapping lobes of T-VII, male.

d. face, male.

e. abdominal pattern, dorsal, male.

f. midtarsus, lateral, male.

g. foretarsus, dorsal, male.

h. abdominal pattern, dorsal, female.

i. S-II, male.

j. S-II, oblique ventral.

k. gonostylus, dorsal.

l. gonostylus, lateral.

m. S-VIII, lateral, male.

n. S-VIII, dorsal, male.

o. mesonotal and metanotal pattern, female.

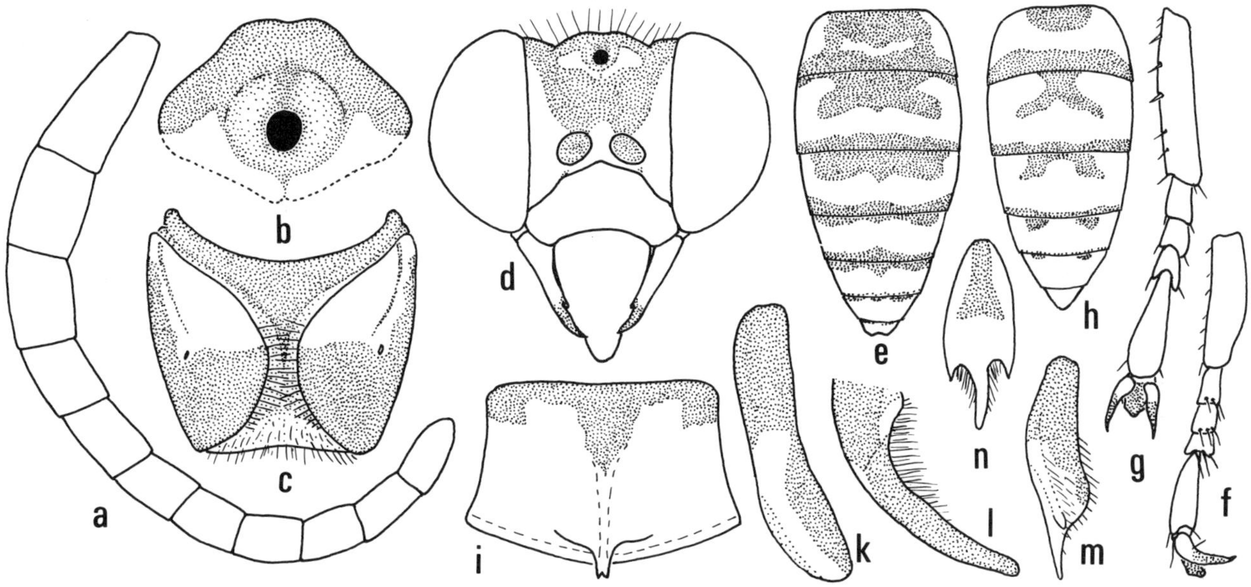

11. megacera

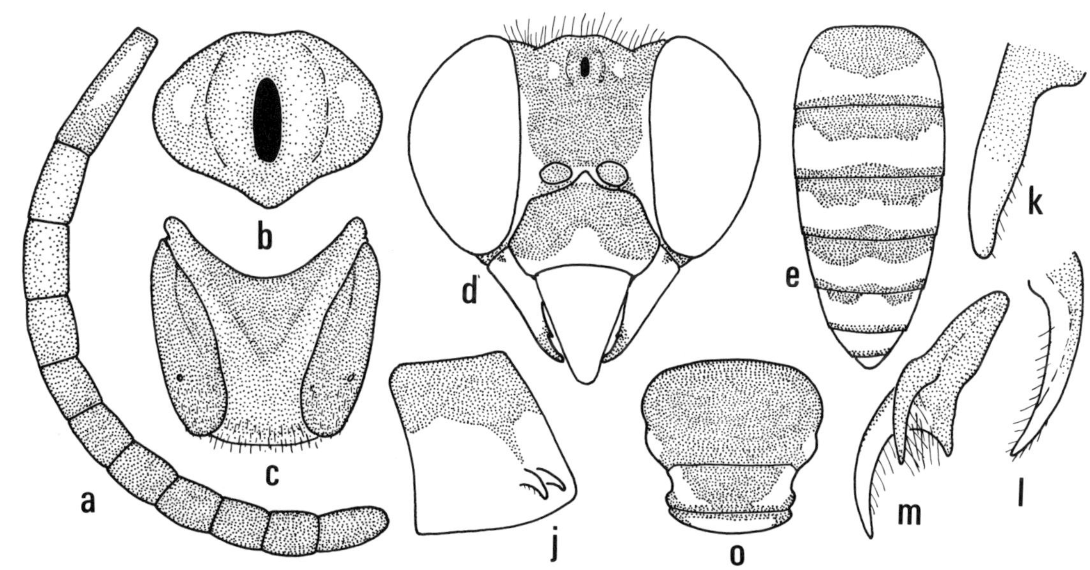

12. pulla

PLATE VII

Figs. 13 and 14. *Glenostictia scitula* (W. Fox) and *Glenostictia tenuicornis* (W. Fox).

a. flagellum, ventral, male.

b. midocellar area, male.

c. S-VII and overlapping lobes of T-VII, male.

d. face, male.

e. abdominal pattern, dorsal, male.

f. midtarsus, lateral, male.

g. foretarsus, dorsal, male.

h. hindfemur, outer lateral, male.

i. abdominal pattern, dorsal, female.

j. head, lateral, male.

k. gonostylus, dorsal.

l. gonostylus, lateral.

m. S-VIII, lateral, male.

n. mesonotal and metanotal pattern, female.

o. midfemur, lateral, male.

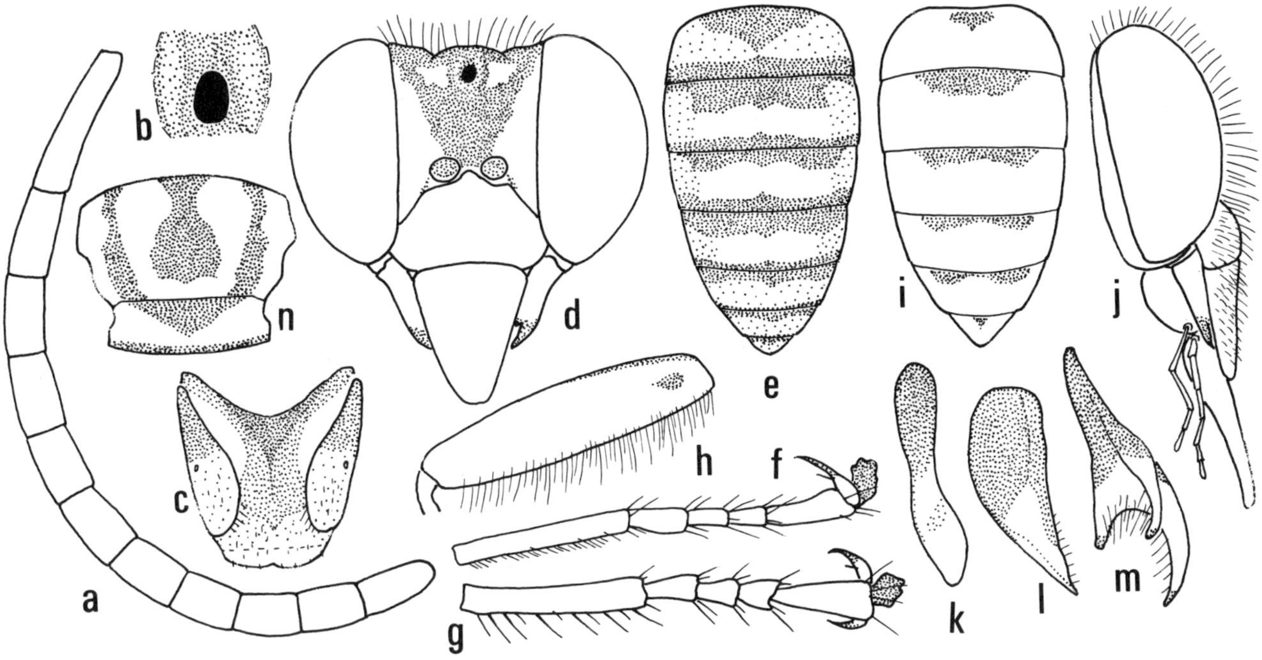

13. scitula

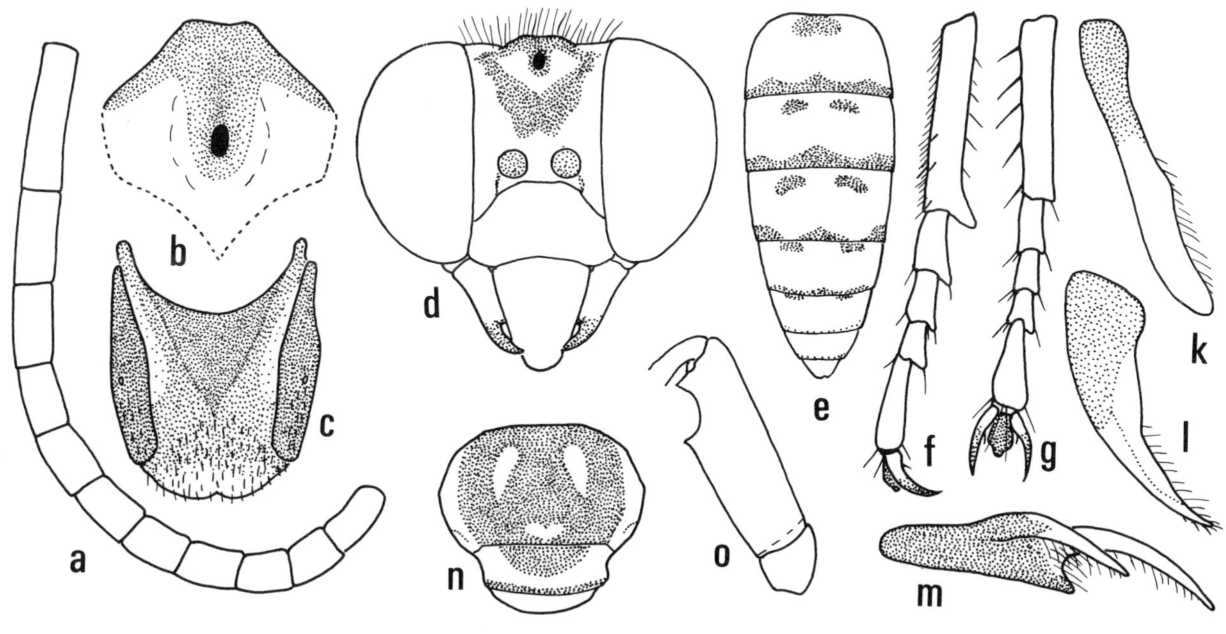

14. tenuicornis

PLATE VIII

Figs. 15 and 16. *Stictiella boharti* Gillaspy and *Stictiella callista* J. Parker.

a. flagellum, ventral, male.

b. midocellar area, male.

c. S-VII and ventral overlapping lobes of T-VII, male.

d. face, male.

e. abdominal pattern, dorsal, male.

f. midtarsus, lateral, male.

g. foretarsus, dorsal, male.

h. femorotibial joint of midleg, posterolateral, female.

i. hindfemur, outer lateral, female.

j. F-IX-XI, dorsal, male.

k. gonostylus, dorsal.

l. gonostylus, lateral.

m. S-VIII, lateral, male.

n. foretarsomeres I and II, dorsal, female.

o. hindtibial apex, lateral, male.

p. mesonotal pattern, female.

q. S-VI, male.

r. S-II, oblique lateral, male.

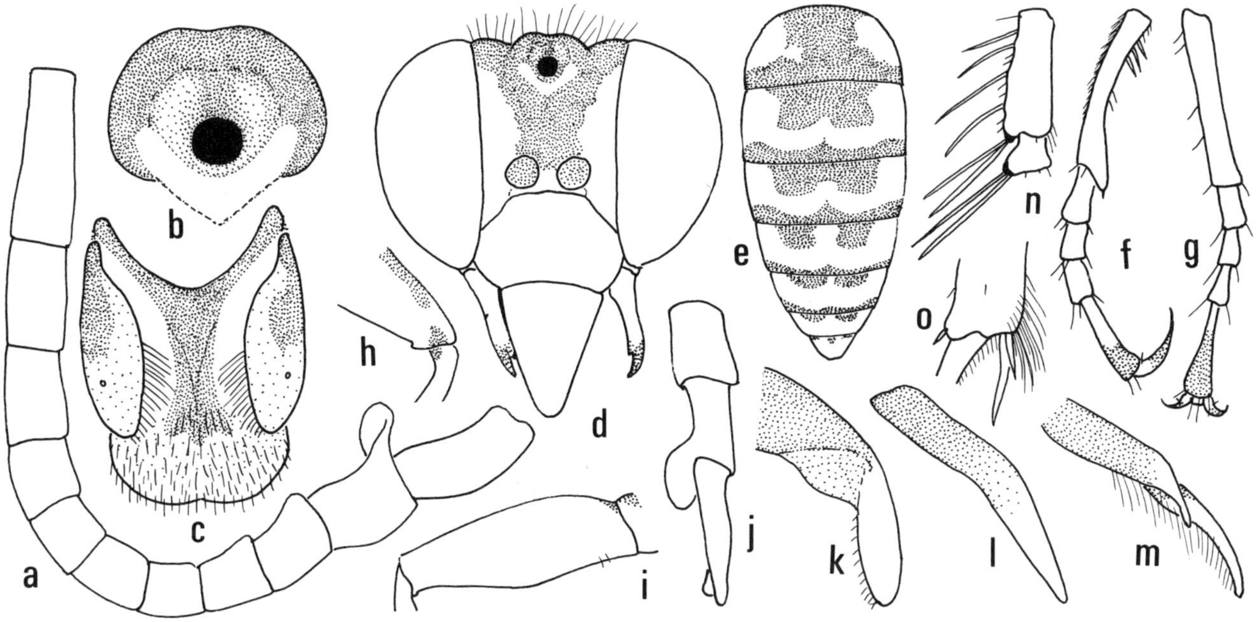

15. boharti

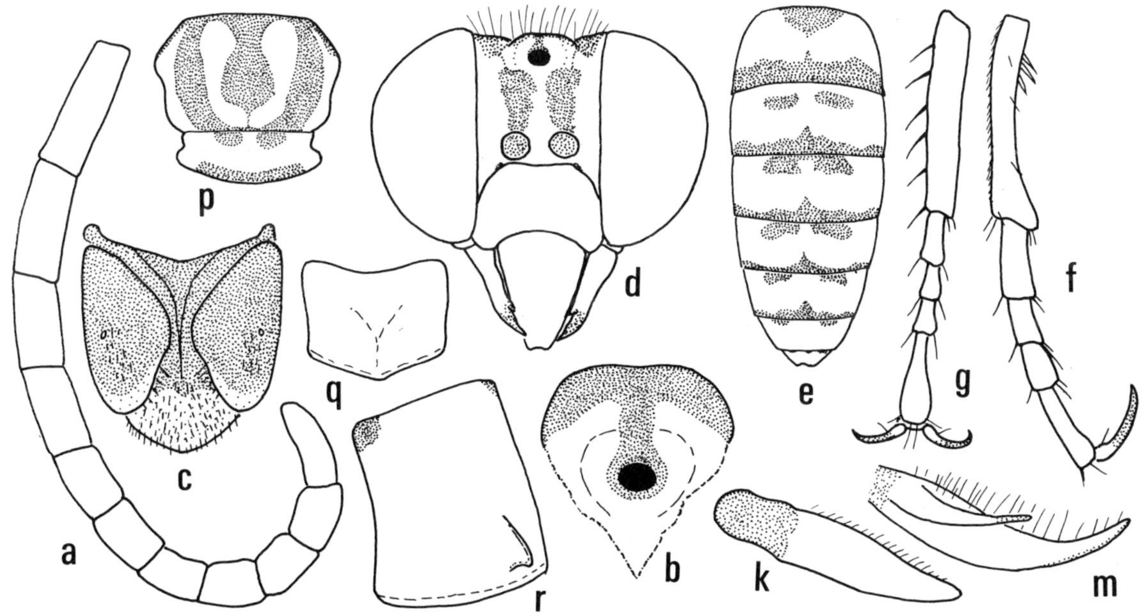

16. callista

PLATE IX

Figs. 17 and 18. *Stictiella corniculata* Mickel and *Stictiella emarginata* (Cresson).

a. flagellum, ventral, male.

b. midocellar area, male.

c. S-VII and ventral overlapping lobes of T-VII, male.

d. face, male.

e. abdominal pattern, dorsal, male.

f. midtibia and midtarsus, lateral, male.

f*. midtarsus, lateral, male.

g. foretarsus, dorsal, male.

h. hindfemur, outer lateral, male.

i. midfemur, outer lateral, male.

j. hindtarsal claws, dorsal, male.

k. gonostylus, dorsal.

l. gonostylus, lateral.

m. S-VIII, lateral, male.

n. S-VIII apex, dorsal, male.

o. S-II, oblique lateral, male.

p. pattern of T-I-III, male variety, Wyoming.

q. pattern of T-I-III, male variety, Riverside Co., Calif.

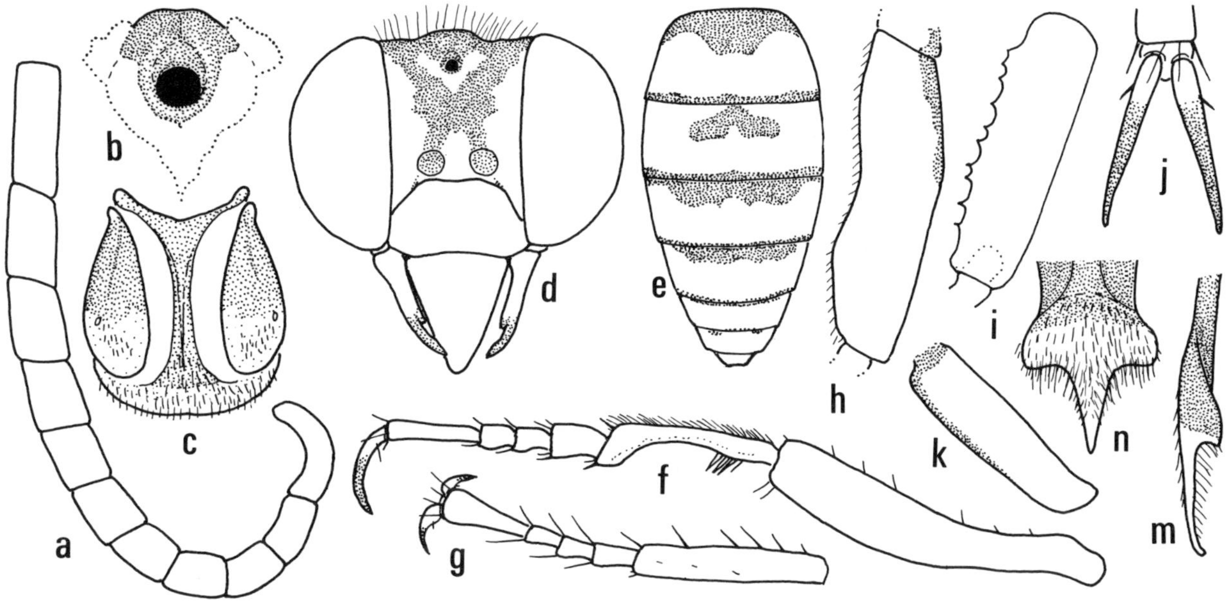

17. corniculata

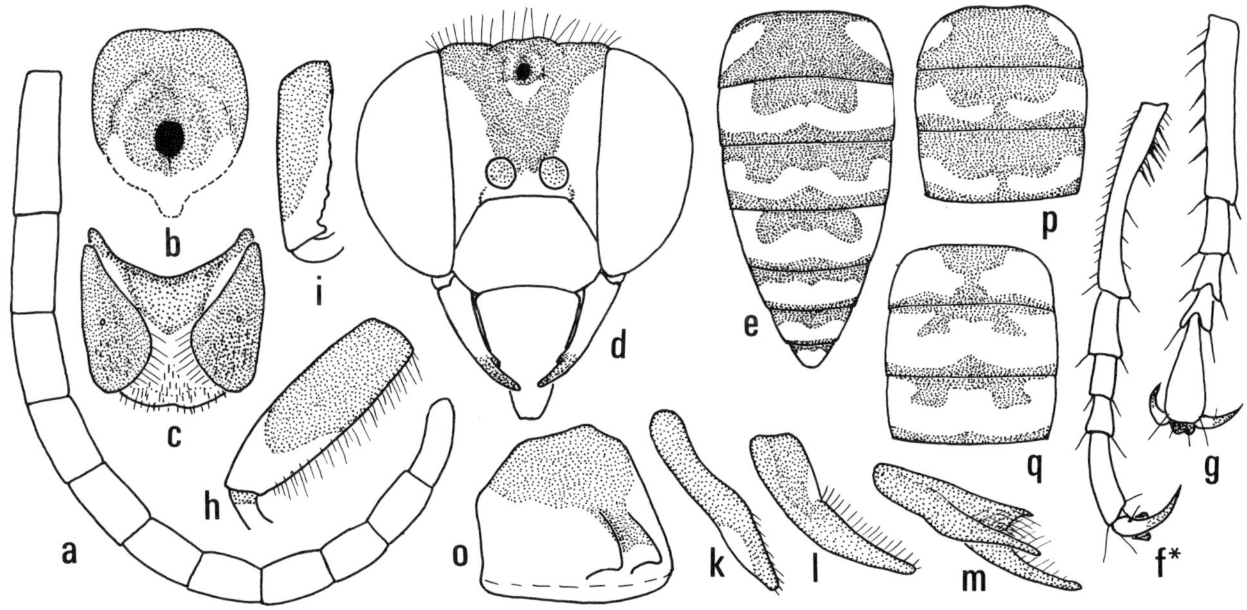

18. emarginata

PLATE X

Figs. 19 and 20. *Stictiella fergusoni* Bohart and *Stictiella flavescens* Gillaspy.

a. flagellum, ventral, male.

b. midocellar area, male.

c. S-VII and ventral overlapping lobes of T-VII, male.

d. face, male.

e. abdominal pattern, dorsal, male.

f. midtarsus, lateral, male.

g. foretarsus, dorsal, male.

h. mesonotal pattern, female.

i. midfemur, outer lateral, male.

j. forebasitarsus, dorsal, female.

k. gonostylus, dorsal.

l. gonostylus, lateral.

m. S-VIII, lateral, male.

n. S-II, male.

o. hindtarsal claws and arolium, female.

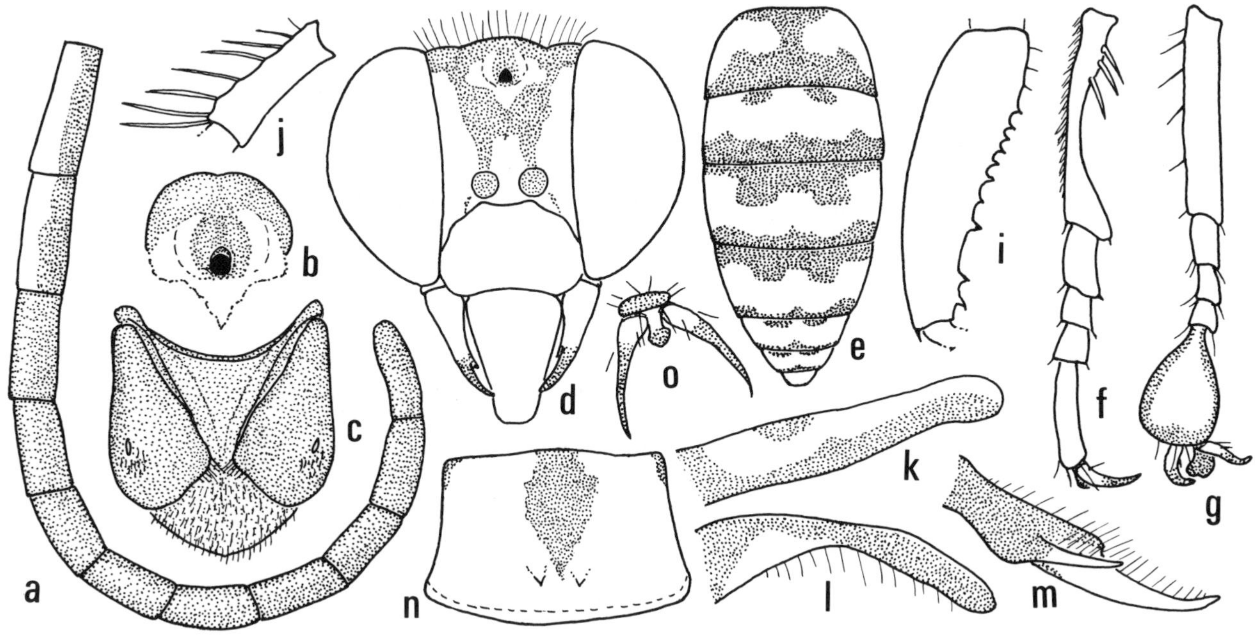

19. flavescens

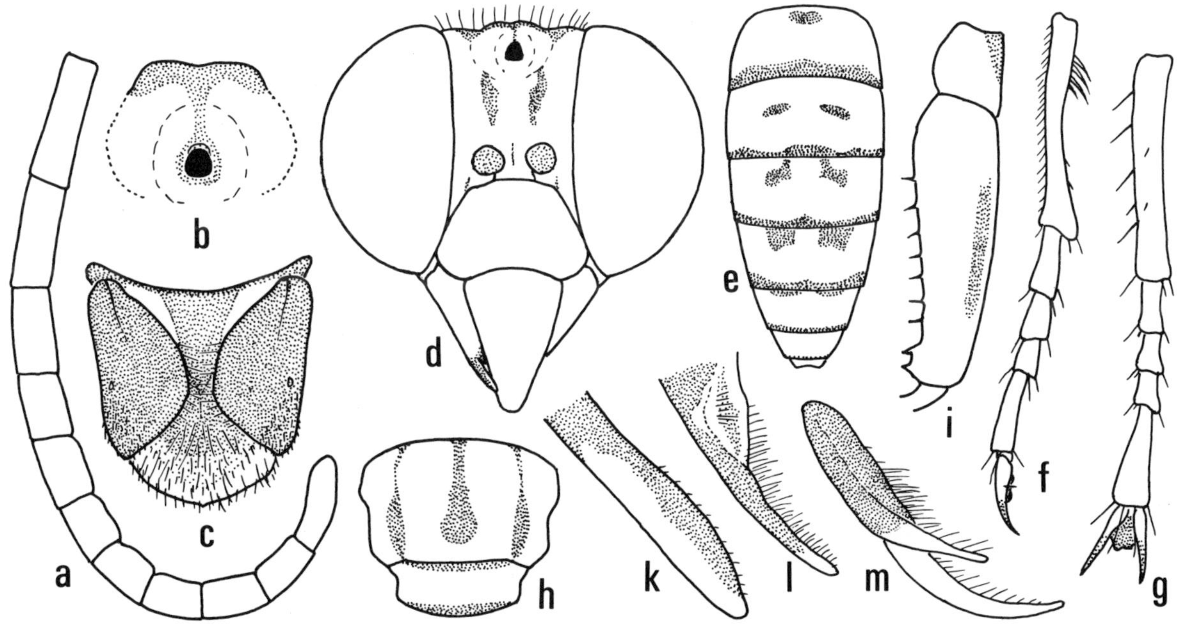

20. fergusoni

PLATE XI

Figs. 21 and 22. *Stictiella pulchella pulchella* (Cresson), and *Stictiella tuberculata* (W. Fox).

a. flagellum, male.

b. midocellar area, male.

c. S-VII and ventral overlapping lobes of T-VII, male.

d. face, male.

e. abdominal pattern, dorsal, male.

f. midtarsus, dorsal, male.

g. foretarsus, dorsal, male.

h. clypeus, female.

i. S-II, oblique lateral, male.

j. mesonotal pattern, female.

k. gonostylus, dorsal.

l. gonostylus, lateral.

m. S-VIII, lateral, male.

n. midfemur, outer lateral, male.

o. clypeus, male variant.

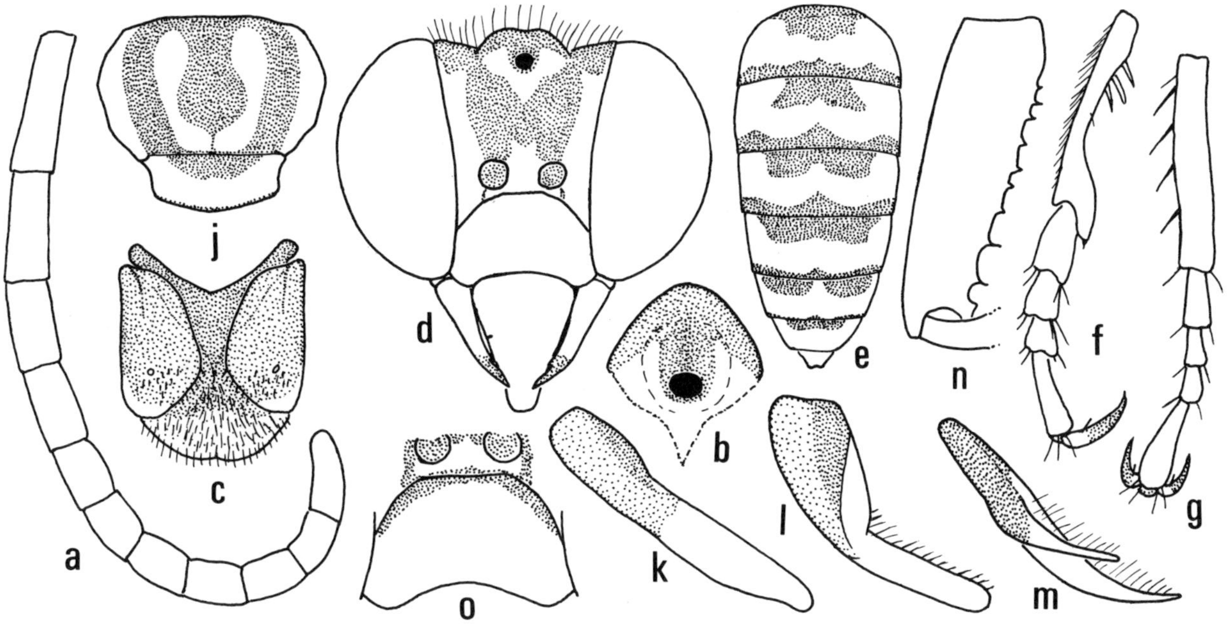

21. pulchella

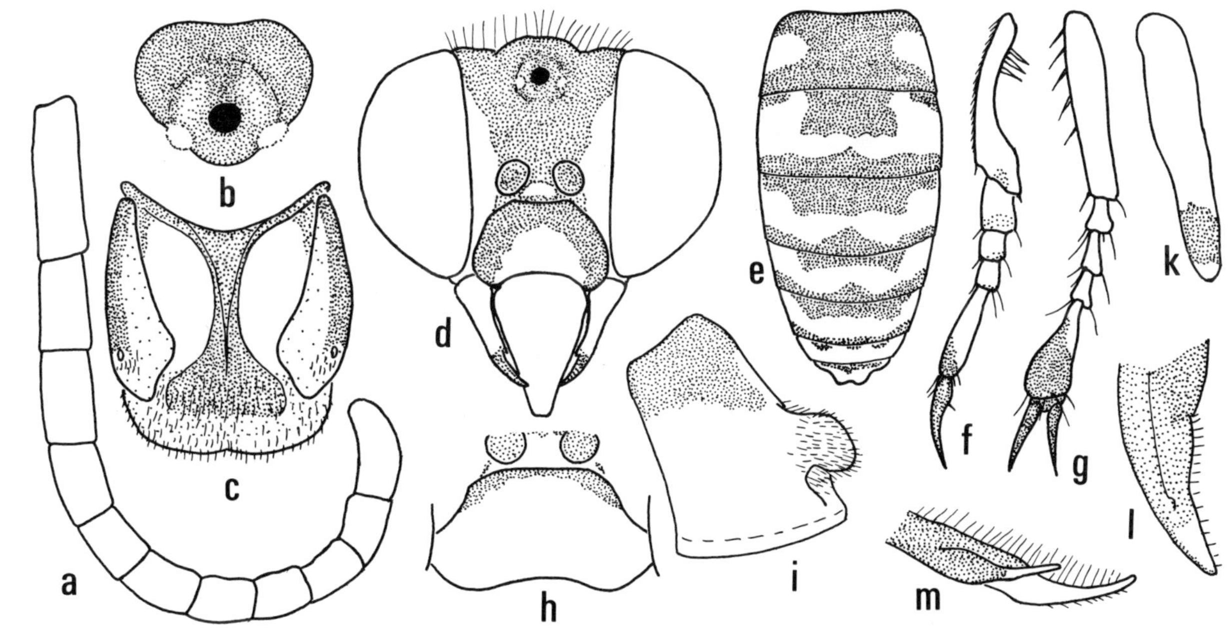

22. tuberculata

PLATE XII

Fig. 23. *Stictiella villegasi* Bohart.

a. flagellum, ventral, male.

b. midocellar area.

c. S-VII and ventral overlapping lobes of T-VII, male.

d. face, male.

e. abdominal pattern, dorsal, male.

f. midtarsus, lateral, male.

g. foretarsus, dorsal, male.

h. mesonotal pattern, female.

i. midfemur, outer lateral, male.

j. S-II, oblique lateral, male.

k. gonostylus, dorsal.

l. gonostylus, lateral.

m. S-VIII, lateral, male.

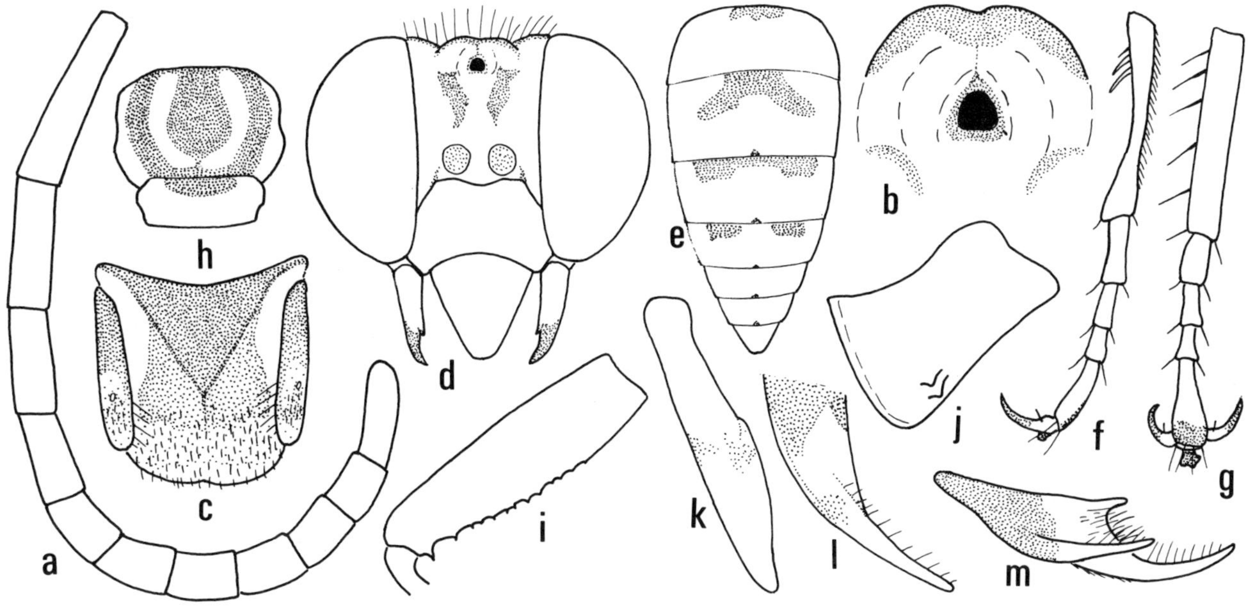

23. villegasi

PLATE XIII

Figs. 24 and 25. *Microstictia deserticola* Bohart and *Microstictia lobata* Bohart.

a. flagellum, ventral, male.

b. midocellar area, male.

c. S-VII and ventral overlapping lobes of T-VII, male.

d. face, male.

e. abdominal pattern, dorsal, male.

f. midtarsus, lateral, male.

g. foretarsus, dorsal, male.

h. hindtarsal claws and arolium, female.

i. head, lateral, male.

j. labial and maxillary palpi, male.

k. gonostylus, dorsal.

l. gonostylus, lateral.

m. S-VIII, lateral, male.

n. S-II, oblique lateral, male.

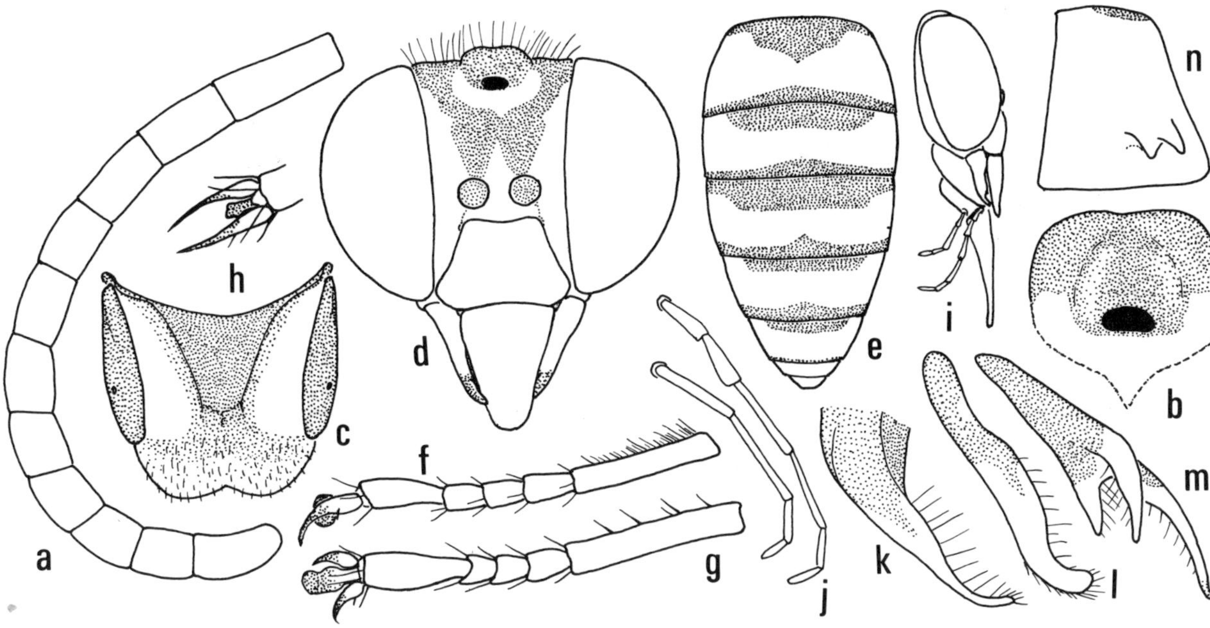

24. deserticola

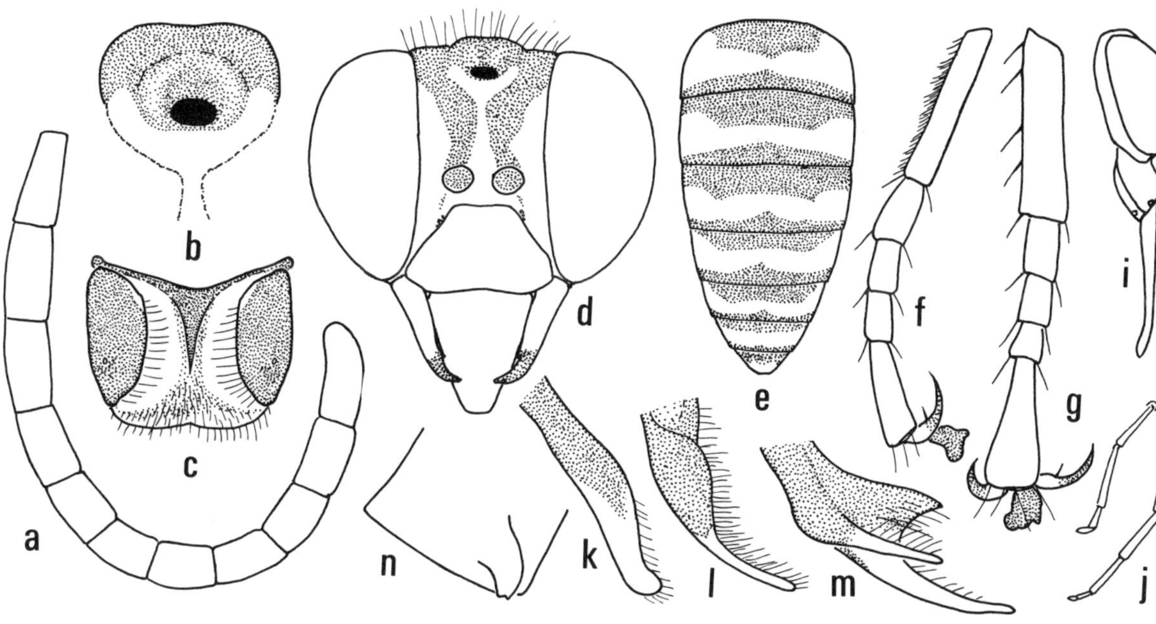

25. lobata

PLATE XIV

Figs. 26 and 27. *Microstictia ocellata* Bohart and *Xerostictia longilabris* Gillaspy.

a. flagellum, ventral, male.

b. midocellar area, male.

c. S-VII and ventral overlapping lobes of T-VII, male.

d. face, male.

e. abdominal pattern, dorsal, male.

f. midtarsus, lateral, male.

g. foretarsus, dorsal, male.

h. S-II, oblique ventral, male.

i. abdominal pattern, dorsal, female.

j. midfemur, outer lateral, male.

k. gonostylus, dorsal.

l. gonostylus, lateral.

m. S-VIII, lateral, male.

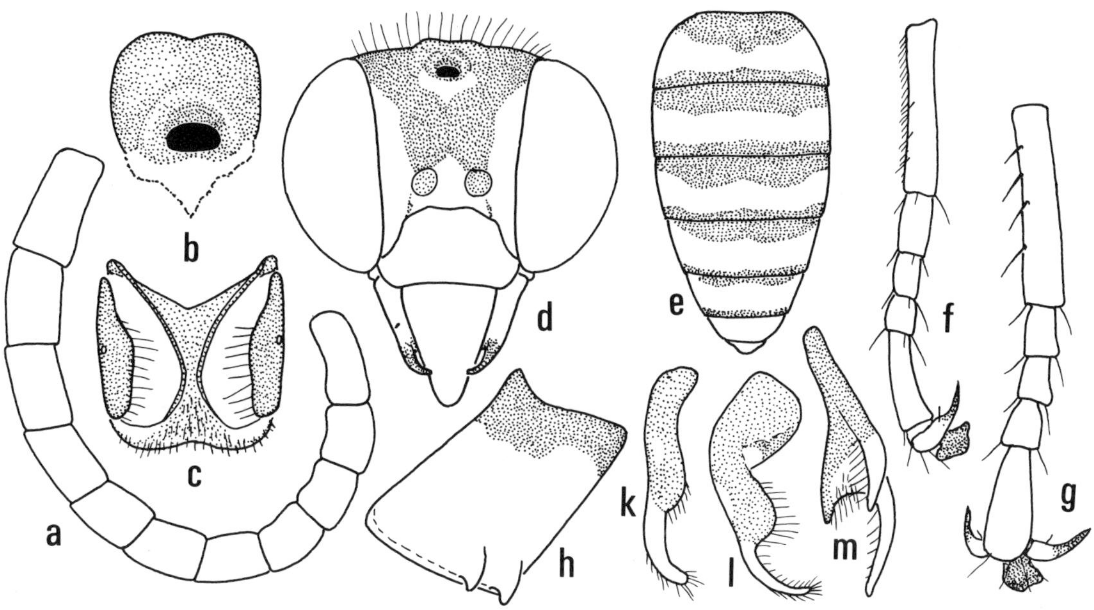

26.ocellata

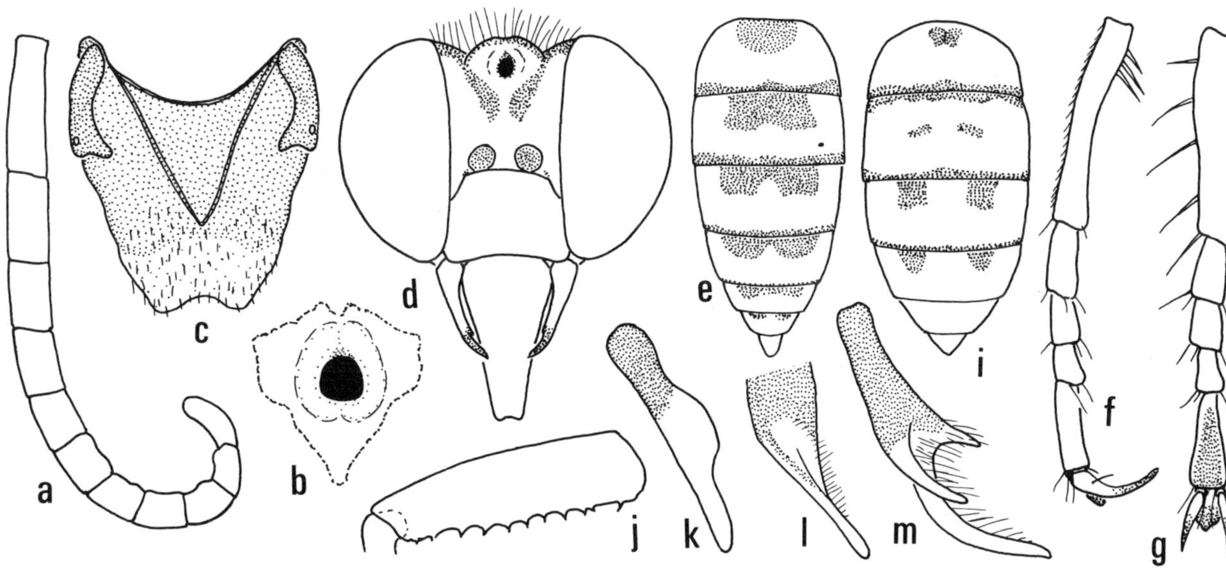

27. longilabris

PLATE XV

Figs. 28 and 29. *Steniolia duplicata* Provancher (28) and miscellaneous *Steniolia* (29): *elegans* J. Parker (eleg.), *nigripes* J. Parker (nigr.), *scolopacea* Handlirsch (scol.), *sulfurea* W. Fox (sulf.), *tibialis* Handlirsch (tibi.), *vanduzeei* Gillaspy (vand.).

a. flagellum, ventral, male.

b. midocellar area, male.

c. S-VII and ventral overlapping lobes of T-VII, male.

d. face, male.

e. abdominal pattern, dorsal, male.

f. midtibia and midtarsus, lateral, male.

g. midtarsals III-V, dorsal, male.

h. F-IX-XI, ventral, male.

i. clypeus and labrum, lateral, male.

j. gonostylus, dorsal.

k. gonostylus, lateral.

l. S-VIII, lateral, male.

m. abdominal pattern, dorsal, female.

n. midtibia, dorsal (outer), female.

o. S-II pattern, female.

p. S-II, oblique lateral, male.

q. S-II base pattern, female.

r. clypeal pattern, male.

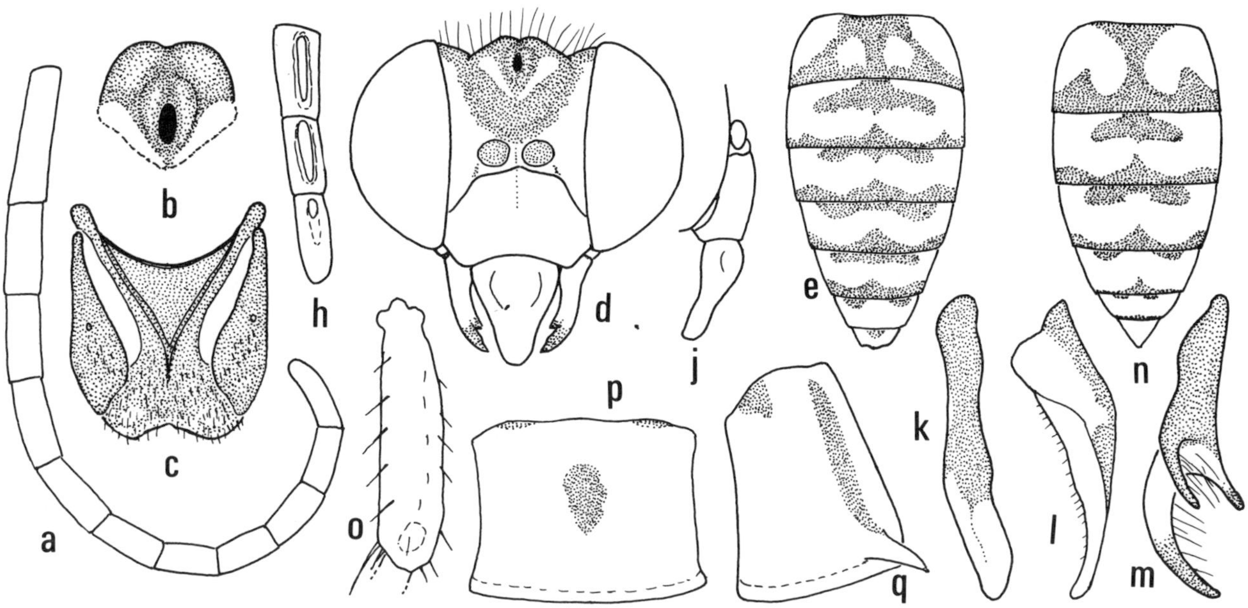

28. duplicata

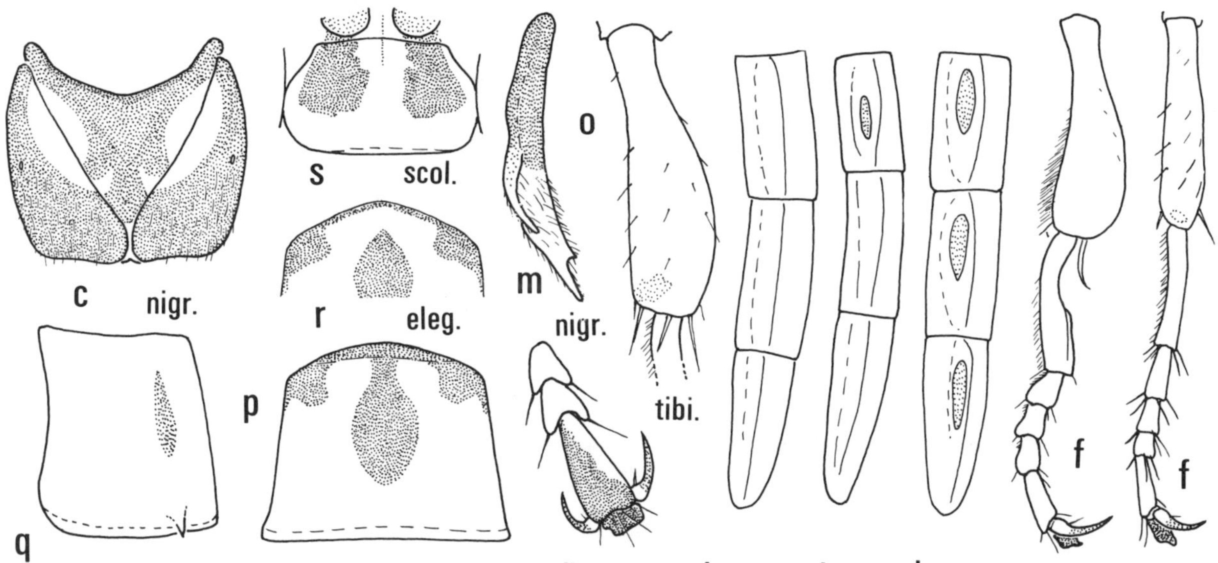

29. Steniolia

PLATE XVI

Fig. 30. Typical T-I-II patterns of various *Steniolia*.

a. *californiensis* Gillaspy, male.

b. *tibialis* Handlirsch, female.

c. *obliqua* (Cresson), male.

d. *sulfurea* W. Fox, male.

e. *sulfurea* W. Fox, female.

f. *nigripes* J. Parker, male, light phase.

g. *nigripes* J. Parker, male, dark phase.

h. *elegans* J. Parker, male.

i. *eremica* Gillaspy, female, light phase.

j. *eremica* Gillaspy, female, dark phase.

k. *scolopacea* Handlirsch, male.

l. *scolopacea* Handlirsch, female.

m. *duplicata* Provancher, male.

n. *duplicata* Provancher, female.

o. *vanduzeei* Gillaspy, male.

p. *vanduzeei* Gillaspy, female.

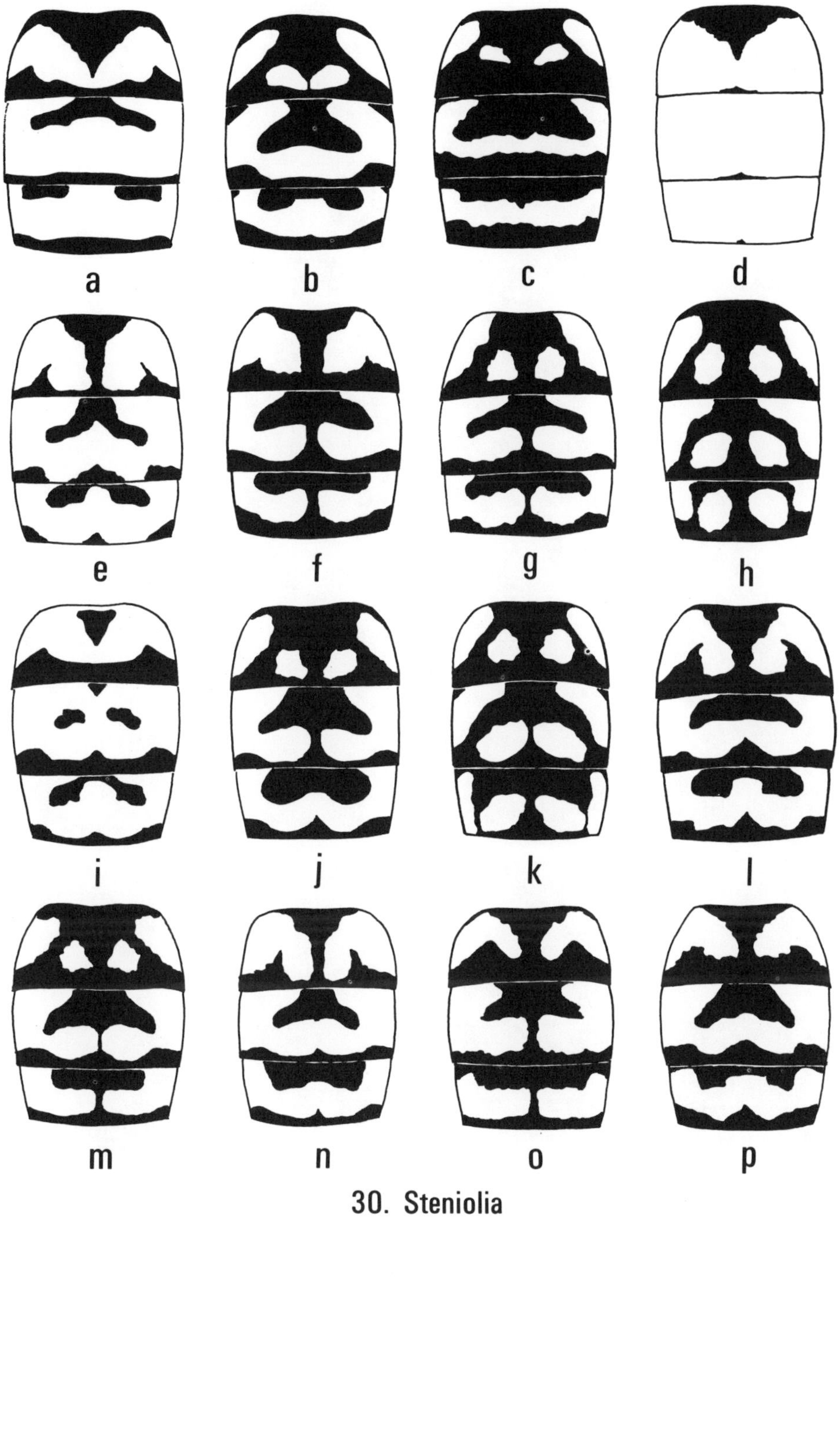

30. Steniolia